BATTLE OF THE GENOMES

The Struggle for Survival in a Microbial World

H.M. Lachman
Albert Einstein College of Medicine
Bronx, New York
USA

Science Publishers

Enfield (NH)   Jersey   Plymouth

The author wants to thank B. Jain Publishers Pvt. Ltd. (New Delhi, India) for permission to use the photograph on the cover of this book, which shows Robert Koch (third from right) on the 1883 expedition to Egypt during which he identified the bacterium that causes cholera.

SCIENCE PUBLISHERS
An Imprint of Edenbridge Ltd., British Isles.
Post Office Box 699
Enfield, New Hampshire 03748
United States of America

Website: http://www.scipub.net

sales@scipub.net (marketing department)
editor@scipub.net (editorial department)
info@scipub.net (for all other enquiries)

Library of Congress Cataloging-in-Publication Data

Lachman, Herbert M.
    Battle of the genomes/H.M. Lachman.
    p.cm.
    Includes bibliographical references and index.
    ISBN 1-57808-432-6
    1. Bacterial genomes. 2. Communicable diseases--Pathogenesis. I. Title.

RB155.5.L33 2006
616.9'0471--dc22

                                                    2006045011

ISBN 1-57808-432-6 [10 digits]
      978-1-57808-432-6 [13 digits]

Published by Science Publishers, Inc. Enfield, NH, USA
An Imprint of Edenbridge Ltd.
Printed in India

*This book is dedicated to my wife Elise, and my children,
Henry and Miranda*

Preface

I was an undergraduate in 1971 when I first experienced the esthetic quality of the relationship between DNA's structure and function, and the elegant way it explained biological phenomena. It filled me with awe. Understanding the workings of DNA provided a bit of harmony and truth to my universe, the way a mathematical proof, piano sonata or religious epiphany might do for others. The defining moment was the solution of DNA's double helical structure by James Watson and Francis Crick in 1953. Their discovery led to the unraveling of the genetic code, the universal plan cells use to convert information encoded in genes into specific proteins, the cornerstone concept that explained the workings of life on earth. It was an intellectual achievement of the highest order. The discovery of the double helix also helped scientists understand the mechanism by which DNA forms a replica of itself prior to cell division, and how mistakes in this copying mechanism, an innate feature of DNA's unique structure and information packaging system, can drive evolution by creating genetic diversity. Every strip of DNA from every living organism carries a unique story of survival. There's also a downside to DNA's error-prone chemistry; it causes birth defects, cancer, and genetic disorders. DNA is also a controversial molecule. Nearly 150 years have passed since the publication of Charles Darwin's *The Origin of Species*, which is really a book about the consequences of DNA variation. Despite the fact that scientists overwhelmingly accept that Darwin's fundamental principle of natural selection -survival of the fittest- guides the evolution of species, including humans, polls show that a majority of people in the United States question its validity. Most of the Darwin controversy concerns *macroevolution*, the grand changes that result in new species formation, which, for many people, conflicts with religious interpretations of the origin of life. But, there's another side to Darwin, *microevolution*, which deals with the consequences of DNA variation on a smaller scale. These changes influence physical or behavioral traits within a species without necessarily changing that species. Pit bulls and Terriers are both dogs, but microevolution through selective breeding has favored genetic changes that create desired differences in their physical appearance and behavior. The fruits and vegetables we eat are modified by microevolutionary genetic forces that are selected by marketing factors; taste, appearance, and shelf-life. Microevolution also plays an important role in day-to-day human health, which, for me, a physician, is the most interesting aspect of Darwin's insight.

Battle of the Genomes specifically chronicles aspects of microevolution revolving around our encounters with catastrophic infectious diseases through the ages -the genetic changes that have resulted in the creation of deadly microbes, and our genome's response to them. One aspect of our response to life-threatening microbes was the development of a multilayered immune defense system, which, among its many weapons, includes a unique method of genetic manipulation that creates a billion different microbe-fighting antibodies. Another has been the acquisition of infectious disease *protective mutations* that have emerged in the human genome over the past ten-thousand years.

However, our antimicrobial defenses do not always work in our favor. The immune system can turn against us and cause autoimmune diseases, such as rheumatoid arthritis or lupus, or allergic reactions, and some protective mutations cause common genetic disorders. It's part of the yin and yang of DNA variation, a recurring theme of this book. The first two chapters of *Battle of the Genomes* present broad overviews of infectious diseases and the immune responses to them. In succeeding chapters, I describe some of the catastrophic infectious diseases that have amassed a genetic arsenal powerful enough to alter the human genome. Interspersed are several chapters focusing on the common genetic disorders these infectious diseases spawned. The last three chapters chronicle a modern twist to the human-microbe battle: scientists have appropriated some of the genetic tools that bacteria and viruses use to increase their virulence to invent recombinant DNA technology. As a result, scientists developed new diagnostic and therapeutic tools to treat disease and fight infections, and, most dramatically, they discovered the genetic basis of cancer.

Battle of the Genomes also considers modern day genetics from a historical perspective. I show a logical line of progression connecting 19th century discoveries to the explosion of scientific findings made in the current genetics era -from Pasteur and Koch, to Watson and Crick. In the fast-paced world of modern research, where publications written only a few years ago may be out of date, the pioneering scientists of the 19th century and early 20th century have too often been neglected. These great figures, whose astute powers of observation made it possible for modern scientists to solve complex medical problems, fight infectious diseases, and crack the genetic code, make a return appearance in this book.

Acknowledgements

The author wishes to thank the family members, friends and colleagues who offered helpful advice and made editorial comments. They are Steve Frankel, Adrienne Hartman, Charles Lachman, Henry Lachman, Benjamin Margolis, Elise Ottenberg, Michael Pollan, and Barbara Rudolph. The author also wants to issue the following disclaimer: This book is not intended as a medical guide to diagnose or treat any diseases.

Contents

Introduction

"An interest in disease and death is only another expression of interest in life" – Thomas Mann

The battle is fierce and invisible.

Everyday, in a universe we cannot see, there is a fight between deadly microbes that surreptitiously enter the human body and our powerful immune system. Humans usually win. If we didn't, we would be routinely overwhelmed by infection from overzealous tooth brushing, which makes the gums bleed a little, or when we cut our skin, circumstances that can deliver a small number of bacteria into the blood. Without an effective immune system, our frequent encounters with cold and flu viruses, and common gastrointestinal pathogens, would turn self-limiting infections into dangerous, life-threatening ones. A handful of powerful microbes, though, have acquired genes that enable them to bypass, evade, or cripple the immune system with such precision that a life-and-death struggle follows. Add on the dimension of contagion, or the spread of infectious microbes by insects, and an infection in one individual can explode into historic epidemics that kill tens of millions of people. These deadly microbes, which include the organisms responsible for malaria, plague, smallpox, tuberculosis, cholera and typhoid fever, and now AIDS, the Biblical plagues, pestilences and poxes, have taken an unimaginable toll on humanity.

Epidemics have erupted throughout the world for millennia, spreading panic, wiping out villages and towns, and in some cases, destroying entire cultures, altering the course of human history in their wake. They also left behind a permanent record of their destructive power by forever changing the human genome. So great was the impact of killer microbes, they created classic scenarios for Darwinian natural selection to occur. Protective mutations emerged in the human genome that worked with the immune system to increase the odds of surviving attacks. These mutations have saved millions of lives through the ages. However, humanity has paid a price. The same mutations that protect some against the ravages of epidemic infectious diseases can cause, under certain circumstances, fatal inherited disorders. Four of the most common of these genetic disorders that share such an intimate relationship with catastrophic infectious diseases are sickle cell anemia, thalassemia, cystic fibrosis, and a strange affliction called *G6PD deficiency* in which red blood cells are destroyed hours after eating fava beans.

Battle of the Genomes chronicles catastrophic epidemics, the genetic disorders they spawned, and attempts by scientists to understand and control both. I also describe how scientists figured out ways to manipulate the very genes infectious microbes use to increase their virulence and resist antibiotics, which led to the invention of recombinant DNA technology and helped to unravel the cause of cancer. The histories of the infectious diseases and genetic disorders described in this book will illustrate the sheer power of DNA. Seemingly simple changes can completely alter microbial virulence, create antibiotic resistance, protect people from infection, cause deadly genetic disorders, and even convert a single normal cell into a giant cancerous growth. When life is viewed through the common lens of dynamic, ever-changing genomes, our relationship with the natural world takes on new meaning and awe. That is powerful stuff for a long string of carbon, hydrogen, nitrogen, oxygen, and phosphorus atoms.

As a clinician and research scientist, I have witnessed every aspect of the battle between microbe and man described throughout this book. In Africa, I saw the listless bodies of small children in the terminal stages of malaria. In New York City, I spent ten years treating patients with sickle cell anemia. In addition, when I decided on a career in research, I learned basic recombinant DNA techniques by studying the beta-globin gene, the gene that provides the link between malaria and several common genetic disorders.

Between the emergence of new infectious microbes and the specter of antibiotic resistance, and the development of common genetic disorders, you will see how Darwinian selection is not merely a concept that helps explain the evolution of species over hundreds of million of years – it is a dynamic force of nature that has an impact on day-to-day human health.

Modern medical science has made it possible for millions of us to live longer and healthier lives than ever before. Most of us plan to work into late middle age or early old age, and enjoy a lengthy retirement. We have reasonable expectations that medical science will solve nagging health problems. Arteries can be unclogged, brains and other organs can be scanned to reveal detailed anatomical features. A cornucopia of medications is available for virtually any condition. Fresh organs are transplanted. Artificial replacements are being developed for the deaf, blind and maimed. There are a few types of leukemia, formerly always fatal, which are now curable. For me though, the most exciting advances have come from the revolution in genetics that began with the discovery of DNA's double helix structure more than 50 years ago, and the ability to treat and prevent infectious diseases. Before the modern era, there was almost nothing people could do to control dangerous infectious diseases. Epidemics struck, like volcanic eruptions, with little or no warning, taking lives indiscriminately. For cholera and plague, death could descend swiftly – not quite like a mountain of falling ash burying you while eating breakfast, but certainly rapidly enough for an encounter with the causative bacteria to kill you in a day or two. Children were especially prone to fatal encounters with infectious microbes. Only half of the children born before 1800 survived their infectious disease-ridden childhoods. The infant-mortality rate in the United States today is about 1 in 200. Childhood infectious disease accounted for 60 percent of all urban deaths in the 19th century. Darwin himself lost his ten year-old daughter Annie to tuberculosis, and blamed himself for passing onto her a predisposition for illness.

Children being weaned are especially prone to serious infections. Nursing infants are provided with partial immunity against infections because breast milk is rich in antibodies that offer temporary protection against bacteria and viruses to which the mother has been exposed. Once infants are weaned, their immune systems must fend for themselves. Throughout human history, and in poor countries of the world today, the seemingly simple and natural transfer from breast milk to whole foods has marked a dangerous rite of passage: weaning has been associated with a startling mortality from infectious diarrhea, as high as 25 percent in some parts of the world, caused by unprotected exposure to a variety of intestinal and respiratory pathogens – a tragic imperfection in human design.

The increased vulnerability of children to epidemic infectious diseases provided the evolutionary driving force for the expansion of the protective mutations described throughout in this book. Survival increased the odds of reaching puberty and passing protective mutations to future generations. So great was the selective pressure that protective mutations, which emerged by chance in a small handful of individuals in the past few thousand years, expanded in the population and are now carried by hundreds of millions of people. Every human living today has some genetic story to tell, if not the protective mutations described in this book, then others, legacies from distant ancestors who were fortunate enough to survive a deadly infection because of unique DNA variations in their genome.

The discovery of the germ theory of disease in the 19th century by Louis Pasteur and Robert Koch, and the development of vaccines, antibiotics, and antiviral medications since then have radically changed the battle against infectious diseases. Today, the modern pharmaceutical industry and a legion of highly skilled academic research scientists searching for new weapons in this battle enable us to have reasonable expectations that serious infectious diseases can be prevented or cured. The major childhood illnesses are nearly extinct thanks to vaccinations. The feelings of anxiety, helplessness, and doom experienced by our ancestors while watching everyone around them succumb to epidemic infectious diseases is hard to imagine today in the developed part of the world, with the exception of AIDS.

Our ability to control so many infectious diseases has even made some of us a little cocky. Conversations about serious bacterial illnesses can be as nonchalant as describing the last movie you saw. I have had several friends casually inform me over the past few years that they or their children were being treated for pneumonia. Before antibiotics became available, cases of bacterial pneumonia were often fatal, and survivors developed chronic respiratory problems from damaged lungs. Nowadays, if we have a sore throat, a visit to the doctor and the corner pharmacy is enough to kill the offending bacterium responsible for the most common serious throat infection, a member of the group A, beta hemolytic *Streptococcus* family. In the past, untreated "strep throat" used to routinely cause kidney failure and rheumatic fever. A sore throat was also one of the first signs of diphtheria, a uniformly fatal childhood infection. Vaccination against diphtheria (the "D" part of the DPT vaccine given to all children in the United States) has effectively eradicated this disease, and those who are infected, can be treated with antibiotics.

We were able to gloat a little over our success in controlling deadly infectious diseases. In fact, in headier times, fresh off the heels of eradicating smallpox and controlling the common childhood illnesses such as measles, polio and diphtheria, public health officials were thinking out loud about the possibility of eliminating all infectious diseases from the face of the Earth. In his address to the U.S. Congress in 1969, William Stewart, the Surgeon General declared, "we can close the book on infectious disease." Sheer hubris!

Since Dr. Stewart's pronouncement, we have witnessed the emergence of HIV, as well as strange and exotic new microbes and infections such as Legionnaire's disease, Ebola virus, dengue fever, Lassa fever, toxic shock syndrome, Lyme disease, West Nile Virus, SARS, and flesh-eating bacteria. Over the past few years, the emergence of antibiotic-resistant tuberculosis and *Staphylococcus*, and HIV resistant to all antiviral medications, reminds us of the stark power of infectious microbes to overcome our most advanced drugs.

In the past, we faced the ravages of smallpox, plague, cholera, tuberculosis, and malaria: today it is HIV, drug-resistant microbes, emerging infections, and, still, malaria and tuberculosis.

Tomorrow, it will be something else. When it comes to infectious diseases, the past is also our present and future for, in the battle for biological supremacy, infectious microorganisms command an unfair advantage over humans: an inherent capacity to adapt to almost any environmental challenge by rapid Darwinian selection, allowing them to reconfigure into new, deadlier versions of themselves in an evolutionary blink of an eye.

CHAPTER 1

The Microbial World

The world has always been, and will always be, a playground for microorganisms. There are millions of different species, a few thousand of which regularly interact with humans. A few hundred of those cause disease, and a relatively small handful have been responsible for catastrophic infections that have resulted in billions of premature deaths in the course of human history.

The diversity of the microbial world on Earth is a marvel. Every drop of water, every speck of soil is teeming with microbial life; microorganisms have found a way to inhabit every conceivable niche in the biosphere, adapting to all of the planet's environmental extremes, from normal oxygen levels to none at all, from the leaves of plants, to the water cores of nuclear reactors and toxic dumps. Some species of bacteria can be found living in habitats below the freezing point of water or above its boiling point. Recently, several species were cultured from an isolated pool of salt water found under an Antarctic ice ridge that had been separated from any other freshwater source for two thousand years – the water temperature was -10 degrees Celsius. The high salt content had kept the temperature well below water's natural freezing point. At the other extreme are thermophilic bacteria, heat-loving microbes that thrive in hot springs and near fissures deep in the ocean floor that vent super-hot, mineral-rich material from the Earth's crust. Two decades ago, scientists at the biotech company Cetus were searching for a variant of the enzyme used by all cells to synthesize DNA, *DNA polymerase*, that would not break down at high temperatures, as most proteins do, to use in the PCR reaction. PCR is the acronym for *polymerase chain reaction*, a Nobel Prize-winning research and clinical laboratory technique invented by Kary Mullis that is used to amplify fragments of DNA. PCR is now the most widespread method for analyzing DNA: its applications range from diagnosing deadly genetic disorders in a fetus inside a womb to DNA-fingerprinting violent criminals. PCR reactions require repeated brief exposures to temperatures above 94 degrees Celsius, which destroys all but the hardiest proteins. Cetus scientists found the heat-stable DNA polymerase they needed, which provided the key missing component required to make the PCR technique a practical research and clinical tool, and which launched a billion-dollar enterprise, by isolating the enzyme that thermophilic bacteria use to replicate their own DNA. The ability of some bacteria known as extremophiles to thrive under extreme environmental conditions has given scientists a glimmer of hope that microbes may exist on other planets and moons in the solar system; Jupiter's moon Europa is the best candidate since liquid water heated by geothermal energy may flow under its surface.

Not content with inhabiting the entire physical Earth, bacteria and other microbes have also managed to carve an existence on the biological Earth as well, in and on the bodies of animals. Thousands of bacterial, fungal and protozoan species have found safe harbor on the skin, in the mouth, between the teeth, and in the intestines of every human being, indeed every animal on the planet, testimony to their extraordinary adaptability. These were first discovered in the 17th century by Anton van Leeuwenhoek, the Dutch inventor of the microscope, who unveiled a world of invisible creatures co-inhabiting our bodies when he observed living microbes, animalcules he called them, in his own dental plaque. "There are more animals living in the scum of the teeth in a man's mouth than there are men in the whole kingdom," he wrote. That was an innocent underestimate. In fact, the total number of individual microorganisms inhabiting a human body numbers in the tens, even hundreds of trillions, equivalent to, and probably exceeding, the number of cells that make up a human being. The combined weight of all the bacteria in any person's gastrointestinal system alone is about two pounds, nearly 2 percent the body weight of a petite woman.

The ubiquity of bacteria in the world provided ammunition for opponents of the germ theory of disease, developed by Louis Pasteur and Robert Koch in the 19th century. Critics argued that such omnipresent creatures could not possibly be responsible for causing disease so indiscriminately. The brilliant experimental work of Pasteur, Koch, and their disciples, discussed in later chapters, proved the critics wrong.

Our microscopic guests, known as commensal microbes, are usually harmless, as long as they stay put in their adopted niches. And, in return for providing a warm and stable, nutrient-rich environment, microorganisms compensate us by providing beneficial services, supplying us with nutrients that we are unable to synthesize on our own, such as vitamin K and vitamin B12. Some gut bacteria may even help in human digestion by activating certain genes found in cells lining the intestinal tract involved in breaking down and absorbing nutrients. Mice grown in a completely germ-free environment gain less weight and are less healthy than mice that live in real-world, bacteria-infested environments.

Perhaps the most important reason for being grateful to bacteria in our gut is that they serve as a protective buffer against unwanted infectious microorganisms. With so many harmless bacteria in the human gastrointestinal tract, there is sometimes too much competition to allow an invading outsider, a pathogenic strain of bacteria, to flourish. This buffering capacity is sometimes reduced when we take a powerful antibiotic, which kills off the protective microorganisms and gives noxious microbes a chance to proliferate and cause disease. For example, as many women come to discover, antibiotic treatment can temporarily reduce the number of normal vaginal bacteria, thereby causing an overgrowth of *Candida albicans,* a yeast, which leads to an irritating infection. Some people augment the normal number of bacteria in their gut by eating live-culture yogurt, which contains the gut-friendly *Lactobacillus* and *Acidophilus* bacteria. There is some evidence that viral diarrhea in children, and vaginal yeast infections in women may be helped by *Lactobacillus* and live culture yogurt. In one study conducted in Kenya, there was even evidence that prostitutes somewhat reduced their chance of contracting HIV and gonorrhea when *Lactobacillus* was introduced into the vagina (a very preliminary finding that one would hope will not encourage the use of yogurt as a prophylactic instead of condoms).

So ubiquitous are the bacteria in our bodies that the only time we and other animals are truly free of them is when we are fetuses in the womb. We begin to swallow these creatures during passage through the birth canal when our mouths encounter vaginal and colonic bacteria, and then during our first suck, be it from a mother's breast or a rubber nipple. By the time children reach the age of two, they have accumulated their own personal mix of microbes that essentially stays with them for life. You can not prevent the intrusion—and you don't want to. People with an extreme view of humanity's routine, everyday exposure to bacteria—who wash, scrub and clean excessively—are viewed as pathological and may be suffering from obsessive—compulsive disorder. Rather than encourage their radical sanitary habits, we send them for psychotherapy and medication to help rid them of their extreme desire for cleanliness.

On the other hand, a too careless attitude toward intestinal bacteria, which can lead to the contamination of food and water, is responsible for about three million deaths every year from infectious diarrhea. This has had a major effect on the human genome, as I will describe in later chapters.

What distinguishes commensal strains of bacteria—those that live in harmony with an animal host-from pathological ones? How do animals tolerate such a gigantic burden of bacteria when the introduction of a few hundred foreign bacteria—potential invading pathogens—can rouse the immune system to action? The microbes that have found safe harbor in humans are genetically adapted for optimal growth in the myriad of environmental challenges we pose, from the hydrochloric acid in the stomach—an acid so strong it can dissolve clothing—to the alkalinity of intestinal secretions, from oily skin to the moistness of mouths. Some acid resistant bacteria, for example, have a gene that codes for an enzyme called *urease,* which produces ammonia, a very strong alkali, from the organic molecule urea, a byproduct of protein metabolism. Ammonia helps bacteria pass through or, in some cases, to colonize the stomach, by creating an alkali-rich microenvironment that locally neutralizes stomach acid. This is one reason for how the bacterium *Helicobacter pylori*, which causes ulcers, thrives in the stomach. The urease gene has made *Helicobacter pylori* one of the most successful bacterial strains in the world; it is found in the stomach of most mammals and billions of people throughout the world.

Bacteria that inhabit the gut are also able to survive the various digestive enzymes present in the stomach and small intestines, without being destroyed themselves, by creating indigestible cell walls or capsules.

Some bacteria that live in the mouth have adapted to life in a sea of saliva, which contains a host of antibacterial proteins, including antibodies, by coalescing as a *biofilm*, a community of bacteria living together as a kind of proto-organism. Like emperor penguins living in the Antarctic whose colony members congregate in a close-knit mass in such a way that the perimeter of birds protect those in the middle from bitterly cold, fierce winds, bacteria on the outer edge of biofilms also help provide insulation for their colony. They secrete a protective coating that prevents saliva from entering the inner sanctum, allowing members of the colony to live peacefully. Their enemies are floss and dental hygienists; dental biofilm is more commonly known as plaque.

Commensal bacteria have also found a way to avoid destruction by the immune system by a variety of strategies, one of which is a clever chemical ruse. The immune system does a remarkable job distinguishing between "self" and foreign (non-self) antigens (antigens are proteins and other substances that stimulate antibody production). The ability to distinguish between self and foreign occurs because immune cells that recognize self antigens kill themselves during development. Otherwise, the immune system would end up attacking normal cells in the body, as it does in autoimmune diseases. Some gut bacteria escape being recognized as foreign by coating their surface with substances that resemble the membranes of host cells, thus fooling the immune system into believing the bacterium is a normal gut component; self instead of a foreign invader.

Being able to adapt to life in an animal without killing it does not necessarily mean that commensal bacteria are completely harmless. The harmony that exists between the microscopic and macroscopic worlds can be breached when bacterial niches change. Skin and mucous membranes provide very tight protective barriers against bacterial penetration and infection. However, when skin is cut, the strain of bacteria known as *Staphylococcus aureus*, which is harmless on the outer surface of skin, can penetrate into the dermal layer underneath and cause a serious infection called cellulitis, or an abscess. If it gets into the blood, the organism can colonize organs of the body, forming abscesses on the brain, liver, and lungs, liquefying large chunks of these vital organs. If it gets into the blood, the organism can colonize organs of the body, forming abscesses on the brain, liver, and lungs, liquefying large chunks of these vital organss. As a medical resident in a large urban hospital in the late 1970s, it was common to see intravenous drug users with *Staphylococcus aureus* abscesses on the lung and brain caused by the bacterium's introduction into the blood stream through unsanitary needles. These are nasty infections indeed, requiring many weeks of intravenous antibiotics. Some addicts did not mind the treatment; the intravenous lines needed for administering the powerful antibiotics also provided easy access for the heroin their friends brought with them during visiting hours.

Most bacteria in the mouth are harmless, but when they leave their usual environment, havoc follows. One of the nastiest skin infections occurs from human bites, which introduce mouth bacteria into the saliva-free, airless confines of the subcutaneous space, where they can grow unchecked.

Harmless bacteria in the intestines can also cause infection when they find themselves in an environment different from their adaptive homes. The common forms of the gut bacteria *Escherichia coli* (known by its abbreviation *E. coli*) rarely cause disease when they stay in the intestines. However, when introduced into the vagina and subsequently into the bladder, a urinary tract infection is sure to follow.

Therefore, the harmless commensal bacteria that share our personal biospheres are really opportunists, Machiavellian schemers, which will take advantage of any weakness occurring in the thin physical barriers that separate them from the blood and internal organs.

Virulent microorganisms, however, do not simply wait for invasion opportunities–they create them. Their survival strategy rests on genetic adaptations that allow them to attack and go for the kill. Virulent microbes have acquired genes that carry instructions for producing toxins and invasion proteins that are used to penetrate protective barriers. Although invasion exposes microbes to attack by the body's immune system, many pathogenic microbes have genetic schemes for evading immune detection either by blocking our first line of defense against penetrating bacteria, the army of bacteria-eating white blood cells (phagocytes) in our blood, or by avoiding the powerful antibody response we generate. Some of these genetic tricks will be discussed later in this chapter and throughout the book.

The conversion of an innocuous microbe into a potential killer is due to the acquisition of virulence factors through gene transfer and genetic variation, intrinsic and inevitable properties of having DNA as life's information-carrying molecule. Although all life forms undergo genetic change and move genes about, microorganisms are true masters.

DNA stores information and transfers it from cell to cell, and from generation to generation in the form of four basic subunits called *nucleotides* or *bases*, adenine, thymine, guanine and cytosine abbreviated A, T, G, and C. Each base is attached to the next forming a continuous structure, like a succession of train cars. Individual genes contain thousands of nucleotides, the precise order of which carries the code for a specific protein. When cells divide, an exact replica of their DNA is made and passed along to both daughter cells. The faithful copying mechanism that doubles a cell's DNA content prior to cell division is called *replication*. Accurate replication is based on DNA's unique double stranded structure. A molecule of DNA is not simply a single continuous series of individual bases, although we write it this way for simplicity. It is actually two strands attached to each other with bases on one strand connected to bases on the other, in pairs. However, there is a rule governing the attachment of two DNA strands, a rule determined by nucleotide chemistry-that is, every A on one strand binds to a T on the other strand, and every G to a C. This A-T, G-C base pairing rule was discovered by Watson and Crick. Prior to DNA replication, enzymes separate the two strands of DNA into individual strands. Then using the enzyme DNA polymerase, nucleotides are added to the single strands. In accordance with the A-T, G-C rule; every A binds to a T and every G to a C. When every blank space on both DNA strands has been filled in with the correct nucleotide, two identical new double strands of DNA are created, each identical to the original double stranded molecule. The genome has been replicated, ready to be partitioned equally into two new cells. This built-in mechanism for accurate DNA replication was forecast by Watson and Crick when they discovered the A:T, G:C base-pairing rule and solved the structure of DNA in 1953. What they saw in their double stranded model of DNA was a perfect structure-function relationship that immediately provided a clue to DNA's copying mechanism. To DNA aficionados, it was one of the great moments of human history, a one-time only epiphany. The A:T, G:C rule also explained a puzzling but key experimental finding made by Erwin Chargaff, which Watson and Crick incorporated into their DNA model. Chargoff had shown a few years earlier that the nucleotide content of every organism was different, but that the amount of A always equaled the amount of T and C always equaled G. Chargoff, who had met Watson and Crick when they were novices at DNA research, and was appalled by their apparent ignorance of basic nucleic acid chemistry, was shocked when he heard the news that they had solved the mystery of DNA's structure.

Some viruses have RNA genomes instead of DNA. RNA has the same information capacity as DNA, and has the same basic structure, except it contains a different sugar component. One other modification-instead of having thymine, RNA has an equivalent base called uracil (abbreviated by the letter U). RNA genomes copy themselves using different enzymes than DNA does, but the copying mechanism uses a base pair matching system similar to the fundamental A:T, G:C rule used by DNA, except that there is a U instead of T on the RNA strand.

The entire complement of an organism's nucleotides constitutes its genome: viral genomes contain tens of thousands of nucleotides and a handful of genes; bacterial genomes contain millions nucleotides with hundreds and thousands of genes; humans have three billion nucleotides and thirty thousand genes. Some viruses, such as retroviruses, have RNA genomes instead of DNA. RNA has the same fundamental structure and information capacity as DNA, with several modifications, which will not be discussed any further. Although the basic structures of DNA and RNA are the same in all life forms, genes and genomes are configured differently in prokaryotic organisms (those without a discrete nucleus to house their genomes–essentially all viruses and bacteria) and eukaryotic organisms (all those with a nucleus–from yeast to human). Genes found in prokaryotes are continuous; they begin and end in a single information strip, which contains the code for a specific protein. In order to convert the information present in a gene into a protein, one strand of DNA is copied (transcribed) into a messenger RNA (mRNA). The order of nucleotide bases present in DNA is precisely copied into mRNA. Messenger RNA is then decoded into a protein through a process called translation, which takes place in a cellular structure called the ribosome–the cell's protein factory. The order of amino acids used to construct a specific protein is determined by the order of bases in mRNA, which, in turn, is determined by the order of nucleotides found in that gene. However, the majority of genes existing in eukaryotic organisms are broken up into discrete gene fragments called *exons* and *introns.* Exons contain nucleotide sequences that code for proteins; introns lack protein-coding information. Initially, RNA is transcribed from the entire gene, copying both the introns and the exons. Then the RNA is processed to remove all the introns, and the exons are joined together to form the final mRNA, which is transported to ribosomes for protein translation. This RNA editing process is analogous to cutting and pasting pieces of film or computer text. In fact, scientists have borrowed a term from film editors and refer to this unique process as *splicing.* The complex of proteins and various RNA molecules required for splicing took perhaps two billion years to evolve, the approximate period of time between the appearance of the first prokaryotic organisms and the first eukaryotic organisms. Why this complicated mechanism of gene activation occurs would be the subject of book on its own. The short explanation is that exons were once primitive genes themselves, and when cells acquired the ability to splice RNA, novel combinations of exons were created to form new proteins that ancestral organisms found quite useful for their survival.

Bacterial chromosomes are more streamlined and gene-packed than eukaryotic chromosomes. Genetic space is such a premium and so crowded with information that genes embedded within other genes are often found; not so for the chromosomes of humans and other higher animals. The exons that make up the 30,000 genes in the human genome represent only a small percentage of the three billion nucleotides in our genetic code. The remainder, the DNA between genes, called *intergenic DNA,* is made up of an odd collection of seemingly useless genetic material. There are pseudogenes, genes that have lost the ability make proteins because they have been bombarded with disruptive mutations, and repetitive sequences, stretches of DNA that have copied themselves over and over again like an out-of-control Xerox machine. There are moveable DNA elements called transposons and retrotransposons that can change their position on chromosomes and disrupt genes, and remnants of ancient retroviruses that attacked the human genome eons ago, and never left. In fact, about 8 percent of the human genome is made up of retroviruses, taking up more space on chromosomes than our genes, as if the human genome invaded a retroviral domain, rather than the other way around. The human genome is a veritable genetic garbage heap, filled with a lot of *junk DNA*, a term coined by Sidney Brenner, a Nobel laureate who helped decipher the genetic code.[1] It is an appropriate metaphor since DNA itself was first discovered in junk: in 1868, a Swiss physician named Johann Friedrich Miescher isolated the chemical basis of life for the first time from white blood cells washed off the pus-soaked, discarded surgical bandages taken from soldiers wounded in the Crimean War.

The tidy flow of genetic information transfer–from DNA to

---

[1] These seemingly useless stretches of DNA are actually not all wasteland. Geneticists are now finding that some elements of junk DNA may have an important function in human evolution. Some of the repetitive sequences help shuttle DNA around the genome to create new genes and gene combinations: and other DNA sequences are transcribed into RNAs that do not get translated into proteins (non-coding RNAs), but instead have a role in regulating the expression of other genes.

mRNA to protein–can be disrupted by mutations. When cells divide in two, the continuous string of nucleotides has to double (replicate) so that the organism's genome can be equally partitioned into two daughter cells. Replication has to be accomplished with a high degree of accuracy so that after a cell divides, the two resulting cells look and behave like the original one-unless they are programmed to do otherwise. However, the process is imperfect. Mistakes are made during replication, approximately once in every million nucleotides. For example, instead of an "A" nucleotide being inserted at a certain position, a "G" or a "C" may be inserted instead. This can alter a gene's code for a protein, which can result in a change its amino acid sequence or affect the amount produced.

On a grand evolutionary scale, making mistakes is a key feature of DNA, and a positive one, in fact. It's the main reason DNA became the information molecule of life. Without an intrinsic, built-in mechanism for creating genetic errors, there would be no variability or change and evolution would stop dead in its tracks. Life on Earth would not have progressed beyond the level of a primitive cell suspended in the primordial soup, replicating exact copies of itself over and over for the Earth's entire existence, or until an environmental challenge wiped out the inflexible organism. However, for us as individuals, there is also a substantial drawback to DNA's inclination for making mistakes. They are responsible for the DNA variations that make us susceptible to most disorders including cancer, genetic diseases, coronary artery disease, obesity, diabetes, schizophrenia and even compulsive gambling, to name but a few.

Genetic variation is also responsible for converting harmless microbial strains into virulent ones, and drug-sensitive pathogenic organisms into drug-resistant monsters. For example, in HIV, mutations in the gene that codes for the enzyme reverse transcriptase (an enzyme that is essential for HIV replication) can cause complete resistance to AZT and other antiretroviral medications in its class, known as reverse transcriptase inhibitors. HIV is a retrovirus–its genome is composed of RNA. Drug resistance develops quite easily in retroviruses because their replication error rate is very high, thousands of times higher than the error rate found in DNA genomes. Retroviral replication is, in a word, sloppy. In people infected with HIV, the viral load (a measure of the number of viral particles in the body) is so enormous and the replication error rate so high, that even before treatment begins, their bodies will harbor a few viruses resistant to AZT and several other antiretroviral drugs because of reverse transcriptase gene mutations that arose by chance. The resistant viruses will selectively grow during AZT treatment, even as the parent virus does not–an inopportune demonstration of Darwinian survival of the fittest. Before multidrug therapy for HIV became available, treatment with only one or two antiretroviral drugs usually led to the emergence of a resistant strain after a few months of treatment. The daily turnover of billions of viruses with a sloppy replication mechanism makes it inevitable.

However, while spontaneous mutations affecting a handful of genes in an infectious virus can alter its behavior, with extremely dire consequences to humans, they are minor compared with the changes that can occur through genetic recombination. Recombination occurs when two stretches of DNA that have similar sequences exchange genetic material. This can lead to the transfer of many genes all at once. Recombination occurs in viruses when a cell is infected with two viral strains that are similar to each other. Flu viruses that affect birds and pigs, for example, can recombine with human influenza virus, creating a novel virulent strain with a unique genetic makeup. Recombination is one of the reasons why different flu shots are needed every year; immunity to last year's strain is ineffective against the new mutant strains that emerge every season in the pig and poultry farms of the Far East. Flu experts and vaccine makers wait anxiously every year for a sneak preview of the influenza strains that have the best chance of spreading around the world during the flu season. They are on the lookout for the lethal combination of genes that created the terrible Spanish flu pandemic of 1918, which killed tens of millions of people around the world, and half a million in the United States, nearly 0.5 percent of the population. Even as this chapter was written, a virulent influenza strain called H5N1 (the bird or avian flu) was ravaging poultry farms in Southeast Asia and millions of chickens and ducks had to be destroyed to stop its spread. Hundreds of people have died so far. Most of these have been poultry farmers: one was a trainer of birds for cockfights. Protecting humans from the transmission of bird influenza viruses is not helped by the unsanitary practice of some cockfight managers who clear the airways of their feathered warriors, ostensibly to improve breathing capacity and increase endurance during battle, by sucking their rooster's nostrils and beak with their mouths.[2]

The bird flu epidemic in Southeast Asia is affecting other mammals besides humans. Nearly one hundred Bengal tigers at the Sriracha Tiger Zoo in Thailand perished after eating raw chickens contaminated with influenza virus.[3] As of October, 2005, the avian flu had spread to poultry farms in Romania, Russia and Turkey.

With the ever-increasing population of the world's people, pigs and chickens, together with the current ease of air travel, and the genetic dexterity of viruses, conditions are riper now than at any other time for dangerous flu combinations to spread rapidly across the globe. But, like forecasting the weather, vaccine manufacturing is not an exact science, and sometimes the vaccines developed in one particular year are not effective against the human flu virus that finally does emerge as the season's dominant strain, like the one that was developed for the 2003-2004 flu season.[4]

Large-scale genetic changes also occur in bacteria through a process called *horizontal gene transfer*, which is the acquisition of genetic material from other microbes. Horizontal gene transfer can swiftly create a brand-new virulent species. One mechanism of horizontal transfer is through *conjugation,* bacterial sex. Yes, even bacteria have sex, wanton, promiscuous sex, in fact, exchanging genetic material with members of their own species, and others as well, by transferring *plasmids*, circular pieces of gene-filled DNA.

---

[2]The Flu Hunters by Gretchen Reynolds, New York Times Magazine, 11/7/04.

[3]Random Samples, edited by Constance Holden. Science 306:808, 2004.

[4]On the other hand, false alarms about dangerous flu strains and the need for widespread vaccination can come back to bite, like the infamous swine flu vaccine in 1976 championed by President Gerald Ford. His very public support for a vaccine, which proved to have serious side effects for hundreds of people, intended to protect against a deadly epidemic that never came, helped cement the impression in the minds of his detractors, along with a few errant golf shots, that the President was inept.

Plasmids are the ultimate parasites, mere circles of DNA with no covering membrane, body or shell, lacking even the bare essentials to qualify as a distinct life form. Although higher organisms required complex evolutionary steps in order to successfully exchange DNA–the development of specialized DNA-carrying cells (eggs and sperm), and the sexually attractive adornments and behaviors that encourage their union, such as ornate feathers, colored rumps, and witty conversation–the process is a bit simpler for bacteria. During conjugation, plasmids are transferred by direct contact from one bacterium to its partner through a pore. The transfer is unidirectional, male to female, only the plasmid-containing "males" express the specialized protein structure on their surface, called the *F-pilus,* through which plasmids move.

Bacteria can also acquire new genes through transposons, movable pieces of DNA, which can enter bacteria by latching onto plasmids, or independently, on their own. They are professional gene-disrupters, moving back and forth between different chromosomal sites, or between chromosomes and plasmids, cutting and pasting DNA along the way, transferring traits from one bacterium to another.

The ability of plasmids and transposons to cross species lines, shuttling genes from one variety of bacteria into another, results in the rapid transfer of traits that improve a bacterial strain's survival or increase its virulence.

The ability of bacteria to procure DNA led to one of the classic genetics experiments of the 20th century, the one that first established DNA as the information molecule of life. Most scientists in the early 20th century believed that proteins carried genetic traits, since their structures are more complex than DNA. It was known that chromosomes were made up of both proteins and DNA, but the DNA portion was regarded as scaffolding, a mere support structure for the more important protein constituent. In 1928, Fred Griffith, an English microbiologist attempted to make a vaccine to protect humans against the most common cause of bacterial pneumonia, *Streptococcus pneumoniae*, but was unsuccessful. However, in the course of his experiments, he found that colonies of *Streptococcus pneumoniae* appeared in culture in two different forms, smooth and rough: bacteria in the smooth colony produce a capsule that helps resist attack by the immune system. Only the bacteria with the smooth capsule killed mice, while the bacteria that form the rough colonies were harmless. He found that if heat-killed smooth bacteria were mixed with the rough strain, the virulence factor could be transferred–the bacteria that make up rough colonies were transformed into killers. Sixteen years later, Oswalt Avery, Malcolm MacCloud and Maclyn McCarty repeated Griffith's experiment, but instead of using whole *Streptococcus*, they separated the bacterium's basic chemical components–DNA, proteins, carbohydrates and lipids–and tested each for the ability to convert the rough strain into the virulent smooth strain. Only the DNA component worked, demonstrating, against popular belief, that DNA, not protein, carried genetic information. This experiment had a profound impact on the scientific world persuading researchers that they had to focus on DNA to understand inheritance. Less than a decade later, Watson and Crick discovered the structure of DNA, and the modern genetics era was born.

The most vexing problem associated with plasmids and transposons is the transfer of genes providing resistance to antibiotics from one strain of bacteria to another. When penicillin first came into use in clinical practice in the 1940's, it was effective against an extensive range of bacteria including all pathogenic strains of *Streptococcus, Staphylococcus*, and the bacteria that cause syphilis, gonorrhea, and meningitis. Very low doses were needed, a fraction of the amount used today. One of the first penicillin recipients during the early clinical trials in 1940 and 1941 was a 43 year-old London policeman named Albert Alexander who was dying of uncontrollable *Staphylococcus* and *Streptococcus* infections in multiple sites throughout the body that began with a cut from a rose thorn he sustained while working in his garden. Penicillin was in such short supply, and the dose needed to control infections was so low, that the drug was extracted from his urine for reuse. However, during the years that followed, the dose of penicillin needed to kill bacteria rose until millions of units had to be prescribed instead of tens of thousands. Eventually complete resistance to penicillin emerged in many strains of bacteria. High resistance to penicillin has been generated by plasmids containing a gene that codes for an enzyme called *beta-lactamase*, which cleaves a portion of the antibiotic, rendering penicillin and many of its chemical derivatives useless. If a single resistant bacterium exists in a population of billions of sensitive bacteria, it will survive a penicillin assault and live to pass down the resistance plasmid to its progeny, and to other strains. A common antibiotic preparation that almost everyone with children has used is Augmentin, which contains the penicillin derivative ampicillin, and a *beta-lactamase* inhibitor capable of restoring some of the antibiotic's killing power against bacteria that have picked up plasmids harboring a *beta-lactamase* gene.

Pharmaceutical companies have developed penicillin derivatives that are impervious to *beta-lactamase*, one of which is methicillin, a powerful antibiotic effective against *Staphylococcus aureus* strains that are resistant to regular penicillin. However, after two decades of use, resistance to methicillin has also emerged in the form of a different plasmid containing a gene called *mecA*, acquired by *Staphylococcus aureus* through conjugation with an unknown bacterial partner. *MecA* carries the instruction for a protein that binds and sequesters methicillin, like a chemical straight jacket, preventing the drug from killing bacteria. Methicillin-resistant *Staphylococcus aureus* is a terrible problem in hospitals, especially in burn units. Until very recently, the only drug effective against the strain was vancomycin, a powerful antibiotic with the high-expectation nickname, "drug of last resort." In 1997, strains of vancomycin-resistant *Staphylococcus* began to appear in hospitals around the world, and soon afterward, a highly resistant strain was found in the United States, in a culture taken from a diabetic with a chronic skin ulcer. Vancomycin resistance is caused by a gene called *vanA*, which jumped species, via plasmid conjugation, when it was transferred to *Staphylococcus* from the bacterium *Enterococcus faecalis* (a gastrointestinal inhabitant, as the name implies). *VanA* is found within a transposon element carried on a plasmid that also houses three other antibiotic resistance genes, plus, for extra measure, a gene that confers resistance to topical disinfectants. Even cleaning hospital floors with disinfectants to reduce microbial growth, a daily ritual in most hospitals (for the health of patients and to gain accreditation), can create selective pressure that allows resistant organisms to thrive. That is why it is better to remove bacteria from the skin using plain soap and water rather than trying to kill them all with antibacterial soaps.

Of all the problems faced by doctors in hospitals, few are as alarming as a report from the microbiology lab of a dangerous bacterium, isolated from a sick patient, which is impervious to every available antibiotic.

Drug companies have recently developed two new antibiotics effective against vancomycin-resistant *Staphylococcus aureus.* However, it's only a matter of time before these too become ineffective. Without the constant development of new drugs capable of counteracting the bacterium's genetic evasive tactics, patients with resistant *Staphylococcus aureus* may have to one day rely solely on their immune system to combat the microbe, or undergo surgery, if the problem is a severe localized infection, a 19th century option for a 21st century problem.

One ancient treatment that some brave and desperate doctors, and their equally brave and desperate patients, have been driven to enlist in the antibiotic resistance era is *larval therapy*–the application of maggots, which is being promoted to treat deep, chronically infected skin ulcers. The maggots, which develop into flies (the greenbottle fly *Lucilia sericata* is a preferred species), feed on the decaying, necrotic parcels found in skin ulcers, cleaning the wound of dead tissue and antibiotic-resistant bacteria. If the patient can tolerate having maggots crawling about, the result after a few days is a clean wound that has a new opportunity to heal. In the few small studies that have compared larval therapy with surgical excision of the wound in curing chronic, infected skin ulcers, the maggots have a slight edge over surgeons.

The problem of antibiotic resistance is an inevitable outcome of bacterial genetics and survival of the fittest. However, the problem has been exacerbated by the reckless use of antibiotics in situations where they are ineffective, such as treatment of viral illnesses. Excessive use of antibiotics in cattle and chickens added to their feed in order to reduce infections and increase growth may also lead to the emergence of drug-resistant strains of bacteria that can spread to humans, such as *E. coli* and several strains of *Salmonella*.

One way to attack the problem of antimicrobial resistance is to use a combination therapy of two or more drugs that have different chemistries and different mechanisms of action, thus reducing the odds that a single mutation will disrupt the action of all the drugs being used. This is why treatment with at least three drugs in combination is now the mainstay for persistent infectious diseases, such as tuberculosis and AIDS. Triple therapy for HIV disease, the antiretroviral cocktail, has made a substantial impact on the health of HIV infected patients. However, because of HIV's rapid mutation rate, Darwinian forces frequently enable the virus to evade extermination, even when three drugs are used, resulting in a need to switch to a new antiretroviral regimen.

The emergence of drug-resistant HIV is not just a problem for the patients already in treatment, but also for healthy members of the population who are vulnerable to transmission of resistant strains through unsafe sex or contaminated needles.

Currently there are enough antiretroviral drugs available so that a different triple therapy combination can be tried and drug resistant strains can usually be stalled at least once. With pharmaceutical companies bringing a new HIV drug to market every couple of years, patients with AIDS have been able to stay ahead of the Darwinian curve, but just barely.

Drug companies have tried to keep pace with HIV partly because of the enormous profits to be made. HIV is an infectious disease affecting tens of millions around the world who will require years, even decades of therapy. The total lifetime cost of treatment per patient is usually hundreds of thousands of dollars. The hundreds of millions of dollars it costs to bring a new drug from development to market is a small fraction of the potential profits to be made, and therefore worth the gamble. There is far less interest in developing new antibiotics that only need to be taken for a few days, the typical treatment time for most infectious diseases.

Pharmaceutical companies also pay more attention to producing drugs to fight the chronic diseases that afflict people living in industrialized countries, such as asthma, depression, osteoarthritis, and hypertension, for which medication is needed for years. The profit stream for these drugs is boundless. Interest in chronic conditions can be seen in the flood of TV and radio ads announcing new medications for joint pain, depression, and erectile dysfunction. When was the last time anyone heard a slick advertisement for a new drug that kills *Staphylococcus?* There is even less financial incentive to create new antimicrobial drugs to treat infectious diseases that primarily affect underdeveloped countries, such as malaria and parasites. Most poor nations can only afford medications that cost pennies for each dose, which is hundreds of times less than the price pharmaceutical companies charge to recoup costs, make a profit, and pay for research and development (and advertising).

The upshot of these economic realities is that completely novel antibiotics that can attack resistant microbes and combat infections are as rare as solar eclipses.

Considering the fact that about 20 percent of new bacterial infections are now caused by strains that have substantial resistance to at least one antibiotic, incurable bacterial infections are bound to become clinical problems unless new drugs are developed. Academic scientists who are less motivated by financial reward, although not totally immune to its seductive power, have the philosophical bent to develop new treatments, but not the resources. For an academic scientist to actually develop a drug on his or her own in a university lab setting, from development at the laboratory bench to clinical trials in patients, would be like building a 747 jet in a garage workshop. Private philanthropic organizations, such as the Bill and Melinda Gates Foundation, are helping to fund the search for unprofitable medications and vaccines, especially those that primarily affect poor nations.

There have also been some promising developments in the scientific front in recent years. The revolution in automated gene sequencing that has resulted in the complete decoding of more than a hundred bacterial strains should provide biotech and pharmaceutical companies with a host of fresh targets for new antibiotic development (more on this in the final chapter).

In addition to antibiotic resistance genes, plasmids also carry other offensive genes that code for proteins capable of converting a harmless bacterium into a killer; these are called *virulence factors*. Non-virulent strains of *E. coli* can become highly pathogenic, for example, after acquiring a large plasmid containing genes that enhance bacterial binding to, and invasion of, the intestinal lining. Another plasmid that has been acquired by *E. coli* contains a gene that codes for a toxin capable of causing dysentery, which results in severe bloody diarrhea. The virulent convert is called enterohemorrhagic *E. coli 0157:H7*, also known as the "burger bug" (for its tendency to contaminate processed meat patties used for commercial hamburgers).

Aside from plasmids and transposons, bacteria can also acquire new genes from bacteriophages, which are viruses that infect bacteria. The life cycle of bacteriophage is a bit more complicated than the plasmid life cycle, mainly because, rather than being just a naked piece of DNA, it is a true organism–a virus–with a protein coat to protect its genome. Bacteriophages latch onto bacteria and inject their DNA. Once inside its host bacterium, bacteriophage DNA directs its own replication, its genes are transcribed and translated to form new viral proteins (using bacterial nucleic acids and amino acids), and newly minted viral particles are assembled. The virus-filled bacterium bursts, spreading infectious particles to other bacteria.

Under some circumstances though, bacteriophages form a partnership with bacteria instead of killing them. Bacteriophage DNA can integrate into bacterial chromosomes where its genes can join forces and become permanent members of the bacterial genome. Instead of surviving by multiplying and killing bacteria, releasing new viral particles to infect other hosts, bacteriophages that integrate into bacterial chromosomes lie dormant, transferring their DNA to the progeny of infected cells. This can sometimes be a more successful survival strategy for the virus than killing bacteria. Chromosomal integration of bacteriophage, also referred to as the latent or *lysogenic* phase, is favored over bacterial destruction when nutrient supplies are low.

Integration of bacteriophage genes into bacterial chromosomes can spell disaster to humans because the virus can shuttle virulence genes from one bacterial strain to another. One example of a virulence factor carried by bacteriophage is the powerful *shiga toxin* gene found in some pathogenic strains of *E. coli*, a poison that prevents protein synthesis in infected cells. Another is the toxin gene found in the bacterium *Vibrio cholerae* responsible for the prodigious diarrhea produced in cholera.

Many pathogenic microorganisms contain blocks of virulence genes called *pathogenicity islands,* which have been introduced into bacterial genomes through recombination with plasmids and transposons, as well as bacteriophages. Pathogenicity islands found in *Salmonella typhi* and *Shigella* species, which cause typhoid fever and dysentery, contain genes that increase gut invasion and induce bloody diarrhea. In addition to the toxin gene, *Vibrio cholerae* expresses a number of other virulence genes transferred en masse into the bacterium thousands of years ago on pathogenicity islands delivered by plasmids.

The pathogenicity island of *Salmonella typhi* contains genes that code for a *type III secretion system*, a complex made up of some two dozen proteins that form a tunnel or channel which bacteria can use to transport virulence factors directly into host cells, like a molecular syringe and needle–visualize two airplanes connected by a refueling line. Three-dimensional imaging of the type III secretion system reveals a structure that resembles an elaborately designed candlestick, one that the most skilled artisan would be proud to fashion. *Salmonella typhi* virulence proteins injected via its type III secretion system disrupt the cellular skeletal matrix, the protein scaffolding that maintains a cell's shape, facilitating bacterial entry and invasion through the intestinal tract. Enteropathogenic *E. coli* is a bit more devious. It uses a type III secretion system to transport its own receptor protein into host cells, which the bacterium can use for attachment and invasion: in the absence of transported receptors, this pathogenic strain of *E. coli* cannot infect cells. It is a bit like sending bomb components to an enemy along with the instructions for construction and self-detonation. The pathogenicity island of *Yersinia pestis*, the bacterium that causes bubonic plague, contains a type III secretion system which is used to transport toxic proteins into bacteria-eating white blood cells, thereby disabling the very cells recruited by the body to destroy the invading microbe. Considering the massive depopulation of Europe caused by the mid-14th century bubonic plague epidemic–the Black Death–which claimed the lives of as many as one out of every three people living in Europe at the time, the chance encounter between an unknown bacteriophage and *Yersinia pestis* thousands of years ago that transferred a type III secretion system to a previously innocuous bacterium, was perhaps the single most catastrophic event in European history, rivaled only by the two 20th century World Wars.

The exposed portion of the type III secretion system needle can be a target for an immune system attack by the invaded host. However, some bacteria counter this by producing an immune-resistant fatty sheath that envelops the needle. Despite this obstacle, scientists are trying to develop vaccines and novel antibiotics that specifically target type III secretion systems, which, if effective, would have a therapeutic effect against a number of different microbes.

The transfer of large numbers of genes by plasmids and bacteriophage combined with the prodigious growth of microbes and the rapid acquisition of naturally occurring mutations can create virulent bacterial and viral strains–and drug resistance–within a few months or years. By comparison, change in the human genome moves at glacial speed, taking decades and centuries to respond to severe selective pressure. With the various genetic options available to microorganisms, the continuous emergence of new infectious microbes and the development of drug resistance are inescapable biological realities.

From the perspective of genetics, it seems as if the deck were stacked against humans. If evolutionary success were judged solely on the speed with which an organism's genome can change and respond to selective pressure and find new growth opportunities, bacteria and viruses would win hands down and higher life forms would never have evolved. The world would be inhabited only by single-celled organisms and their viral parasites–which in fact was the extent of the first two billion years of life on Earth. The emergence of complex organisms and their success in a microbial world was contingent, along with many other adaptations, on the development of an effective immune defense strategy against infectious invaders that could compete with the speed of genetic variation and Darwinian selection that occurs in infectious organisms, a feat accomplished, as discussed in the next chapter, by the body's antibody producing cells.

CHAPTER 2

Magic Bullets

How do higher animals cope in a world dominated by infectious microbes? Similar to the way humans handle life-threatening situations, from sending men and women into space to maintaining a hospital's flow of electricity during a power outage, the defense against infectious microbes requires a multi-layered approach, with built-in redundancies and backup systems. This did not happen all at once, and the defense against infectious organisms did not begin with the emergence of complex animals. In fact, the very first bacteria needed to defend themselves against the viruses that were emerging on the ancient Earth, as will be described in chapter 14.

One of the earliest defense strategies adopted by complex multicellular organisms was to endow subsets of cells, called *phagocytes*, with the ability to engulf and destroy invading microbes. The most abundant phagocyte in blood is called the neutrophil, which, along with other phagocytic cells, such as macrophages, are produced in the bone marrow. A tiny drop of blood the diameter of a pencil point contains thousands of them. Phagocytes are cellular mercenaries looking for action. They dart through the blood, lie in wait along the inner lining of blood vessels, and enter tissue, searching for bacteria and fungi to devour. Billions enter and leave the blood every hour in pursuit of their microscopic prey. Bacterial walls and chemical factors released at the site of an injury attract phagocytic cells to help mop up microbes before they have a chance to proliferate. Phagocytes follow the direction of the chemical scent, like a bloodhound on assignment, using membrane appendages called pseudopods to propel them towards the source of the signal. When contact is made with an errant microbe, phagocytic cells wrap and engulf them with a membrane envelope, and destroy the prisoner with enzymes and chemicals.

Phagocytes were discovered in the 19th century by the Russian embryologist and developmental biologist Ilya (Elie) Metchnikoff (1845-1916). He was a brilliant student with a photographic memory who began publishing scientific papers as a teenager. Tempestuous and grandiose, Metchnikoff also experienced severe bouts of depression (symptoms suggestive of bipolar disorder). Metchnikoff discovered phagocytes in 1882 while working in a makeshift lab he set up in Messina, Italy. He had traveled to the Mediterranean to escape political turmoil in Russia and to recuperate from a severe depressive episode, during which he attempted suicide by injecting himself with a dose of the spirochete bacterium that causes the tick-borne illness, *relapsing fever*. His great discovery was made while studying digestion in sea anemones and starfish larvae taken from the warm Mediterranean waters. Metchnikoff had added a colored dye to his microscopic preparation of tiny anemones as a visual aid. Unexpectedly, he observed motile cells in the anemone surround and devour the dye. He suspected that the cells were eating the dye as a form of defense. He also observed the same type of cell surround a thorn which he had implanted in a starfish larva–the thorn had been plucked from a tangerine tree his children were using for Christmas. The cells resembled amebas and white blood cells that he had previously observed in pus. He then injected bacteria into larvae and witnessed their destruction by the ameboid-appearing cells. He called the cells phagocytes after the Greek words *phagos* (to eat), and *cyte* (cell). His discovery laid the foundation for understanding the host defense against the microorganisms that were being discovered as the cause of infectious diseases by Pasteur and Koch. The discovery of phagocytosis was one of most important biological achievements of the 19th century. It earned Metchnikoff a share of the 1908 Nobel Prize in Physiology or Medicine with Paul Ehrlich, who had done fundamental research on antibodies. The shared Nobel Prize was an appropriate recognition of Nature's dual approach to fighting infections–phagocytosis and antibodies–although Metchnikoff and Ehrlich feuded about the relative importance of each.

The discovery of phagocytes also saved Metchnikoff's life. He stopped suffering from suicidal depressions and became obsessed with preserving health. Good bowel hygiene and eating yogurt were his health mantras. He died of heart disease at the age of 71, a relatively long life for a man at that time, although Metchnikoff had expected that his healthy habits would keep him going for a hundred years.

Aside from phagocytes, complex organisms can counter microbial invasions with antibacterial proteins that interfere with bacterial growth or, in some cases, actually kill them. The antibacterial protein mucin, for example, which is secreted by glands in the mucous membranes lining the respiratory, gastrointestinal and urinary tracts, attract bacteria like fly paper, preventing them from penetrating underlying tissue. Another protein, lactoferrin, limits bacterial growth by sequestering iron, which bacteria need for some of their metabolic pathways. Alexander Fleming discovered the first antibacterial protein, lysozyme, which is found in tears and nasal mucus, a decade before his discovery of penicillin. Fleming was actually more excited about lysozyme's prospects for clinical use than he was about penicillin. However, studies showed that it acted more against harmless bacteria than harmful ones.

Perhaps the most important antibacterial substances produced in animals, after antibodies, are the serum proteins called *complement*, which can stimulate phagocytes and directly kill bacteria by punching holes in their cell walls.

Another important defense mechanism against infectious microbes involves a family of membrane-bound proteins called Toll. Toll proteins were first discovered by a German scientist, Christiane Nusslein-Volhard, in the late 1980's in the fruit fly, *Drosophila melanogaster*. Dr. Nusslein-Volhard named the protein after the German word for weird because flies that lack Toll had developmental abnormalities that made them look odd. However, a role in protecting against infection was suspected when it was found that the lack of Toll also resulted in a terrible fungal infection that covered the fly's body like a fur coat.

Soon thereafter, a family of related proteins, called Toll-like receptors (TLRs), was discovered in higher organisms. TLRs coat the surface of white blood cells and other cell types, sniffing out and sensing the presence of invading microbes. They are found throughout the animal kingdom (and in the kingdom of plants as well)–nearly a dozen different TLR genes exist in the human genome. Each TLR has the capacity to recognize and attach to a specific chemical component of bacteria and viruses. The interaction between a TLR and its chemical trigger activates a number of antimicrobial pathways, such as phagocytosis, and the production of antibacterial proteins and cytokines, a family of proteins that rouse the immune system into action and stimulate inflammation.

The interaction between a TLR and its chemical target is quite specific. Toll-like receptor 4 (TLR4), for example, binds to lipopolysaccharide (LPS), a complex carbohydrate bound to a lipid, found on the surface of certain bacteria (the so-called gram negative variety, which includes *E. coli*). Another TLR, TLR5, is activated by flagellin, a protein found in the flagellum, the whip-like tail motile bacteria use to propel themselves. Because so many pathogenic bacteria have LPS in their cell walls, or use flagella to move about, the capacity of TLRs to recognize these vital structures and stimulate several defense systems constitutes a simple strategy for animals and plants to respond to a wide array of infectious microbes.

There can be a downside to TLRs, however. The cytokines produced by TLR activation are a properly balanced brew of substances that both stimulate and constrain immune and inflammatory responses–failure to keep the body's response to infectious microbes in check can lead to serious problems of their own. For example, a too vigorous reaction triggered by the interaction between TLRs and bacteria contributes to the redness seen in severe cases of acne, the bloody diarrhea found in inflammatory bowel disease, and septic shock (collapse of the circulatory system that occurs with certain infectious microbes). Bacteria with LPS in their cell walls injected into laboratory mice cause septic shock and death in a few hours; LPS alone has the same effect. TLR4, and the inflammatory response it elicits, is the culprit. Mice with defective TLR4 genes are completely resistant to the effects of LPS.

These defense strategies–antibacterial proteins, TLRs, and phagocytes–constitute elements of what is called the *innate immune system*, the earliest, most primitive forms of defense against microbes that emerged through evolution. By primitive, I mean primordial not crude; innate immunity is a very effective defense strategy. Insects rely on it entirely to counter potential pathogens while living in some very bacteria-rich environments such as dung and decaying flesh. However, innate immunity can be overcome by sophisticated microbial trickery. *Mycobacterium tuberculosis*, which causes tuberculosis, is able to survive very long periods of time, decades in many cases, by living inside macrophages, avoiding destruction by recruiting a host protein called TACO, which blocks the ability of macrophages to deliver their cargo of bacteria-destroying enzymes to the invader. The ability to protect itself from being digested by macrophages is one reason for the success of *Mycobacterium tuberculosis* through the ages. Tuberculosis has endured for thousands of years. It has been found in ancient bones, and in mummies from the Andes and Egypt. Despite the availability of effective medications over the last 60 years, *Mycobacterium tuberculosis* still infects nearly half of the human population.

Viruses have also developed elaborate strategies for evading the innate immune system. For example, humans and other higher animals have a gene with the robotic sounding name APOBEC3G, which codes for a protein whose function it is to edit DNA and RNA, altering the genetic code of certain genes in a positive way to generate useful, modified proteins. These gene editors also double as an innate anti-viral defense system by inhibiting retroviruses, such as HIV. It accomplishes this by chemically altering the retroviral genome, essentially scrambling its genetic code. HIV has learned how to counter this effect using a retroviral gene called *vif,* which codes for a protein that blocks APOBEC3G. Some scientists working on novel methods to treat HIV infection are trying to develop drugs that neutralize *vif,* which would restore the natural antiviral action of ABOBEC3G in HIV-infected cells.

An important drawback of innate immunity is its inability to remember a previous attack by a microbe–i.e, when that microbe invades an organism again, the same responses are generated and there is little or no significant immune priming. Immune memory that would help counter subsequent attacks by the same or a similar microbe, especially those that produce a very high mortality rate, would provide a big evolutionary advantage. Another limitation is that innate immunity relies on the recognition of a relatively small number of microbial components, LPS for example. Many pathogenic organisms have learned how to cripple major arms of the innate immune system with relatively simple genetic modifications, such as using *vif* to bind ABOBEC3G. Imagine a less selective system in which any microbial component, not just one or two, can trigger a host immune reaction. Microbes would require sophisticated genetic tinkering in order to escape such a defense strategy.

The disadvantages of innate immunity were addressed nearly half a billion years ago when the adaptive (or acquired) immune system emerged in cartilaginous fish, which provided vertebrates with the ability to generate the most versatile antimicrobial proteins in nature: antibodies.

Each antibody produced in the body has the capacity to bind to specific foreign substances, particularly microbial proteins and polysaccharides. Cells called B-lymphocytes, which are found in lymph nodes, blood, and throughout the intestinal and respiratory tracts, produce antibodies. Its companion microbe-fighter is the T-lymphocyte, which targets intracellular infectious microbes, those that try to escape immune system surveillance by hiding inside of host cells. T-lymphocytes also have anti-cancer properties and attack transplanted organs (the drugs that transplant recipients take to prevent organ rejection are designed to limit the action of T-lymphocytes).

The main improvements offered by adaptive immunity over the innate immune system are the capacity to respond to an almost limitless myriad of foreign substances with specifically engineered, designer antibodies and T-lymphocytes, and a built-in memory feature. Immunologists refer to the capacity lymphocytes have to generate so many different antibodies (approximately a billion) as GOD, an acronym for "generation of diversity".

One of the early pioneers in antibody research was Ehrlich. As a young researcher, he distinguished himself by developing a staining procedure, a modified version of which is still used today, to help doctors identify *Mycobacterium tuberculosis*, which Robert Koch had just discovered. Ehrlich got a bit too close to his research, literally, and contracted tuberculosis, forcing him to take the "rest cure" in Egypt. After two years, he returned to Europe, and secured a position in Koch's Institute for Infectious Diseases, where he teamed up with Emil von Behring and Shibasaburo Kitasato, a Japanese microbiologist, to develop an antiserum treatment for diphtheria (Kitasato is also known for his controversial role in the discovery of the plague bacillus, discussed in chapter 12).[5] The unique specificity of the antitoxin-toxin interaction Ehrlich studied helped solidify his belief that chemical compounds could be generated or extracted from nature that would specifically attack receptor targets on microbes: he called these compounds "magic bullets." Ehrlich was nearly a century ahead of his time. Finding specific molecular solutions to medical problems lies at the very foundation of modern biomedical research. Specificity translates into more effective weapons with fewer side effects, although it does not always pan out that way (AZT was believed to be very specific since it is an inhibitor of an enzyme only found in retroviruses, but it can cause a host of unpleasant side effects).

Of all the magic bullets in nature, antibodies are the most sophisticated and marvelous; their synthesis is based on a unique genetic prescription. Antigens, usually proteins and polysac-charides (complex carbohydrates) produced by bacteria, viruses and other infectious agents, are the prime intended targets (antigen is the term used to describe a foreign substance that can

---

[5]Antiserum is produced by injecting large mammals—horses were usually used—with sub-lethal doses of a bacterium or toxin. Animals thus treated generate large quantities of antibodies against the bacterium or toxin, which scientists can harvest from serum, the liquid portion of blood.

generate antibody production). However, antibody production is not solely directed against microbes. They are generated against proteins found in other people. That's why blood transfusions and transplanted organs have to be matched as closely as possible to the recipient. Some women even produce antibodies to proteins present on their sex partner's sperm. Antibodies can also be produced against substances found on non-living materials, such as latex. The immune system is the ultimate biological xenophobe; it dislikes all things foreign.

Antibodies are complex proteins composed of a large *heavy chain* and a smaller *light chain*, each of which consists of two parts: a *constant region,* which is similar for all antibodies of a given type, and a *variable region*, which is different for every antibody. Light chains bind to heavy chains, creating a complex that binds to itself to form the complete antibody, a protein with a Y-shaped configuration. The stem of the Y contains the heavy chain's constant region and the Y's divergent lines contain the variable regions of both the light chain and the heavy chain. The structure of antibodies was elucidated by Gerald Edelman and Rodney Porter who shared a Nobel Prize in 1972.

There are five basic classes of antibodies produced by B-lymphocytes, defined by the type of heavy chain constant region attached to the stem of the "Y". Each one has a specific role in the immune response. The IgM class, for example, is expressed early in the defense against infections, whereas IgG is expressed later, and for a greater duration, sometimes for an entire lifetime. This type of partitioning has an important survival feature. IgM antibodies have five antigen binding sites for every one found on IgG, so it is a better microbial attacker at the beginning of an infection. However, IgG has a longer half-life, making it a more durable defender against future attacks.

The IgA class of antibodies is secreted in breast milk, providing protection to nursing babies. It is also found in the gastrointestinal and respiratory tracts, providing a first line of defense against ingested and inhaled pathogens.

The IgE antibody type helps protect against parasitic infections, a useful function in underdeveloped countries, but one that is outweighed in industrialized nations, where parasites are rare, by its other role; IgE mediates allergic reactions. The antihistamine medications allergy sufferers regularly consume block the actions of the chemical histamine, which is released from a few cell types found in the blood and in the mucous membranes in response to an IgE-bound antigen. IgE is such a potent inducer of histamine release that a person with a peanut allergy can develop a severe reaction simply by kissing the lips of someone who has just eaten a peanut butter and jelly sandwich. In addition, as people with an allergy to dust are so annoyingly aware, IgE can even induce the release of histamine and other allergy modulators in response to proteins present, of all places, in the feces of dust mites.

The least abundant class of antibodies is IgD, which sits on the surface of B-lymphocytes. When IgD binds to antigens, it stimulates the proliferation of its resident B-lymphocyte, which then produces IgM and IgG antibodies against the same antigen.

The variable region is where antibody binds to the antigen. It is what accounts for the extraordinary ability humans and other vertebrates have to mount a response to virtually every conceivable foreign antigen. It is so versatile that B-lymphocytes have evolved with the capacity to generate antibodies against microbes present at the dawn of vertebrate evolution, which are now extinct, as well as to newly emerging microbes. In fact, we fully expect the same cells to produce highly specific antibodies to microbes that may not evolve until millions of years in the future, as well as extraterrestrial microbes, should we ever encounter them. Unfortunately for the aliens in the book and film "The War of the Worlds," and for indigenous population isolates decimated by diseases carried by Europeans–to which they were exposed for the first time during the age of exploration–a robust immune system is sometimes not enough to fight off a brand-new infectious microbe.

The built-in adaptability of the immune system is similar to some aspects of brain function in that both evolved with the capacity to respond to environmental challenges that would not emerge for millions of years. The motor system of the brain, for example, developed with the ability to coordinate movements not present at the dawn of man, such as riding a skateboard, hitting a baseball, driving a car through traffic, or playing piano. The memory pathways that allowed us to recall an excellent hunting area, a fresh water supply, dangerous animals or plants, or what our mate looks like, also gave us the capacity to memorize batting averages, rules of algebra and commercial jingles.

However, there is an apparent mathematical and biological impossibility to the extraordinary plasticity of the antibody-generating system. Since antibodies are proteins and proteins are made from genes, how can a billion different antibodies be created when there are only approximately 30,000 genes in the entire human genome? Even if every bit of junk DNA coded for antibodies, there is just not enough to make so many different proteins. Evolution's solution to the problem of generating antibody diversity from a limited amount of DNA is to build them using combinations of gene fragments or modules, randomly mixing and matching smaller numbers of them to build up larger constructs. Three sets of genes, for example, each containing a 1,000 different modules, would generate a billion different combinations, if they randomly assorted with each other (1,000 times 1,000 times 1,000).

Creating genetic diversity from a small number of gene fragments can be visualized by a menu analogy. If a restaurant makes five different pastas and ten different sauces, it can actually make fifty different dishes by utilizing every possible pasta and sauce combination (5 times 10). Provide the option of sprinkling on 5 different cheeses and you have 250 different variations. Instead of needing a ten-page menu to list every possible pasta dish, everything can be put onto a single page. Just choose one item from columns "A" (pasta), "B" (sauce), and "C" (cheese). Antibody diversity is generated the same way.

The genes that instruct lymphocytes to synthesize light chains are found on chromosomes 2 and 22: the genes for the heavy chains are on chromosome 14. The light and heavy chain loci contain hundreds of different gene fragments. During B-lymphocyte development, the fragments that form an antibody's variable region randomly rearrange to produce a unique genetic combination that codes for a single protein containing the variable region of both the light chain and heavy chain, both of which combine with a constant region gene fragment to form the gene sequences needed to create a complete antibody. Because the process is random, each individual B-lymphocyte has a unique combination of rearranged light chain and heavy chain gene fragments which, after transcription and translation, produces an antibody that is unique to that particular B-lymphocyte. Once rearrangement takes place, it is permanent, except for a few modifications, one of which is described a little later in this chapter. Extending the menu analogy a little, one B-lymphocyte is programmed to make ziti with marinara sauce and Romano cheese; another is programmed for spaghetti with meat sauce and Parmesan cheese. Even though each lymphocyte has the potential to prepare any combination of pasta, sauce, and cheese, it is only programmed to make one dish, which it does over and over again.

The shuffling or recombination of variable region gene fragments to create a unique antibody gene combination only occurs in B-lymphocytes, even though the gene fragments are found in every cell in the body.

If you calculate the total number of different permutations created by the recombination of variable region gene fragments, a billion different possible combinations are produced, representing the entire repertoire of different antibodies one can generate in a lifetime. This extraordinary mechanism for creating antibody diversity was discovered two decades ago by Susumu Tonegawa, a Japanese immunologist, who was awarded a Nobel Prize in 1987.

Gene shuffling and recombination is also the mechanism used to create diversity within T-lymphocytes. T-lymphocytes do not produce antibodies, but instead, have receptors on their cell membrane that are capable of latching onto a foreign antigen. Each T-lymphocyte manufactures a different antigen-binding receptor, and, analogous to antibody formation, receptor diversity is generated by randomly shuffling receptor gene fragments to create countless possible combinations.

The extraordinary capacity antibody and T-lymphocyte receptor genes have for generating diversity, which is so critical for our survival in the microbial world, can sometimes have dangerous consequences. Mistakes happen. Instead of linking together different antibody gene fragments, the region can inadvertently recombine with genes that have nothing to do with immune function. If this happens to a gene involved in regulating cell growth, inappropriate stimulation of cell division can occur, which can lead to cancerous transformation. This is how many cancers of the bone marrow and lymph nodes–lymphomas and leukemias get a jump-start on their path towards unstoppable growth.

The antibody response to foreign antigens is very selective; a B-lymphocyte is activated when the IgD antibody expressed on its surface membrane latches tightly to a foreign antigen. This interaction triggers cell division and the production of large quantities of the specific antibody synthesized by the activated B-lymphocyte. Out of a pool of hundreds of billions of different B-lymphocytes in the body, only a handful produces antibodies that will bind to a particular invading microbe. These will proliferate, producing a small legion of antibody-producing cells, each armed with millions of antibody molecules. All the other B-lymphocytes remain quiescent, awaiting the arrival of their specific microbial targets in future infections. The overwhelming majority of them will never be called into action. The same selective activation exists for T-lymphocytes; only those with receptors that bind tightly to antigens proliferate during infections.

This selective response to microbes is critical for human health and survival. Not only does it conserve resources, the simultaneous activation of many T- and B-lymphocytes is downright dangerous, initiating a cascade of inflammatory reactions capable of causing damage to the blood vessels and internal organs severe enough to kill a person.

The activation of a subset of B-lymphocytes producing the most appropriate antibody against a microbe does not begin and end with their initial encounter. Stimulated B-lymphocytes are capable of producing even more effective microbe-binding antibodies by modifying their unique variable region through a process called *hypermutation*. The mutation rate in the antibody forming genes of stimulated B-lymphocytes is about a million times greater than other parts of the genome, and it results in small modifications of the primary amino acid sequence of the antigen-binding antibody. If the modification leads to an antibody with an even greater antigen-binding capacity than the original, it will selectively latch onto microbial antigens and proliferate, and large quantities of the improved antibody will be secreted. Hypermutation ensures the high-level production of antibodies with absolutely the best possible fit to foreign antigens that nature can supply. Repeating the pasta analogy one last time, a B-lymphocyte that is only programmed to make spaghetti with meat sauce and Parmesan cheese, can be improved by switching from commercial grated Parmesan in a jar to a fresh imported variety. It is the same dish (original antibody) but tastes better (modified antibody generated after hypermutation).

Antibody-coated microbes are recognized and destroyed by phagocytic cells more efficiently than uncoated microbes. Complement proteins are also activated by microbes that are coated with antibodies, unleashing their bacteria-killing power. Thus, the innate and adaptive immune systems cooperate and communicate with each other, illustrating a general evolutionary engineering strategy: when new biological systems are assembled, pre-existing ones are integrated into the design.

The selective growth of a subset of B-lymphocytes capable of binding to specific antigens, from a pre-existing pool of billions of B-lymphocytes, is referred to as clonal selection. It is essentially a form of Darwinian selection occurring in the immune cells of the body; antigen-stimulated cells have a growth advantage and multiply, those that are not stimulated remain quiescent. Proliferative advantage is also afforded to T-lymphocytes whose T-cell receptors provide the best specific fit to an antigen.

Clonal selection was predicted in the late 1950s by the great Australian immunologist Sir Frank MacFarlane Burnet (1899-1985), and Niels Jerne (1911-1994), a Danish investigator, many years before the structure of antibodies and antibody-coding genes were elucidated. Although the idea was arguably Burnet's greatest contribution to immunology, he won a Nobel Prize in 1960 for a different discovery: immune tolerance, the process by which the immune system inactivates self-recognizing antibodies during development. Without immune tolerance, the immune system would attack the very body that houses it, as patients with autoimmune diseases are so painfully aware.

Although the ability of the adaptive immune system to respond to a wide array of foreign antigens is remarkable, it is not perfect. The forces of the microbial world can still overwhelm us: otherwise, we would never get sick or suffer infant mortality. In fact, the highly efficient system of antibody production developed in the course of vertebrate evolution has a serious flaw: clonal selection and efficient antibody synthesis against a new infectious microbe takes a day or two to fully mobilize. While the immune system is revving up antibody production and activating T-lymphocytes for an attack, infectious microorganisms capable of doubling their numbers every 15 to 30 minutes can accumulate, overwhelm innate immunity, and cause illness. Only after the adaptive immune system has had time to mount its highly specific response against the offending microbe can recovery occur. The delay in the adaptive immune response when a new microbe appears on the scene is one of the reasons for the awesome destructive power of never-before seen infectious microbes, the terrible effects of which have been demonstrated countless times in human history. Witness the effects of bubonic plague when introduced for the first time in Europe during the Middle Ages, and measles and smallpox brought by Europeans to indigenous populations in the Americas and Polynesia.

The immune response is faster once an individual has survived a previous exposure to an infectious microbe, or has been provided with a vaccine, exploiting the immune system's memory feature. The response is faster the second time around because the previous attack by a living microbe, or antigen exposure through vaccination, leave behind large numbers of B-lymphocytes which act as *memory cells*, already primed with the capacity to produce microbe-specific, hypermutated antibodies. All they need is the stimulus of microbial antigens to begin proliferating and producing antibodies again. In addition, residual IgG and IgA antibodies present in the blood, the respiratory system and the gastrointestinal tract, produced during the prior exposure, immediately begin to sequester and destroy a returning invader.

The adaptive immune system can be fooled by sophisticated infectious microorganisms that have evolved with the capacity to evade host antibodies and T-lymphocyte responses. One way they do this is by switching the production of essential surface proteins from one subtype to another, which is like changing the outer candy covering of an M&M candy from brown to red; it is the same recognizable entity, but with one slightly different characteristic. Just when the immune system has established its specific antibody attack against an invader, a critical antigenic determinant (the brown M&M) on the microorganism's membrane may switch to such an extent that the binding site on the antibody no longer fits the altered antigen (red M&M). After the switch, the host has to mount an immune counterattack against the new antigen, triggering additional antigenic switches. The upshot of an infection by an organism with the capacity to change surface proteins is chronic illness, gradual immune victory, or death.

An under-appreciated key player in the battle with infectious microbes is the human brain, which interacts at many levels with the immune system. Receptors for cytokines, for example, are found throughout the brain where they regulate fever, pain, memory, and the release of hormones critical for responding to emotional and physical stress. The brain, through connections it makes with the sympathetic nervous system–the collection of nerves that causes the release of adrenaline and makes animals defend themselves or run or when confronted with danger–also primes the immune system by increasing the numbers of phagocytes and lymphocytes in the blood. However, it is our intellectual capacity and our ability to acquire and pass down knowledge to succeeding generations that has really turned the brain into an organ of the immune system. Before the modern scientific era, human beings were virtually on the same footing as any other member of the animal kingdom afflicted with infectious diseases, aside from medicinal herbs available in some environments, and the comforting effect of praying. With a few exceptions, it was only the emergence of scientific reason and experimentation and the advent of the Industrial Revolution that provided humans with the means to understand the nature of infectious disease and construct effective weapons against it. However, it took quite a bit of time for human beings to reach the point where we could use our brains as a weapon against infectious disease. The science and technology needed to understand and control infections came much later than other comparable human abilities. Ancient humans built monumental structures, such as the pyramids in Egypt and the Great Wall of China. Yet, at the same time these architectural wonders were conceived, there was widespread belief that disease was caused by imbalances in yin and yang (according to the precepts of the so-called Yellow Emperor) or in the four humors (according to the ancient Greeks). In Renaissance Europe, painters and composers were creating timeless combinations of form, color, sound, and allegory. However, people living in that time believed that epidemics were caused by sinful behavior, the poisoning of wells by Jews, and vapors emitted from rotting organic matter. In England, while Shakespeare was composing English language poetry and prose that still endure after three centuries, contemporary doctors bled and purged sick people, destroying any vestige of health remaining in their severely ill patients and reducing the chance of spontaneous recovery. Even something as fundamental as hand washing was not recognized as an infection preventive measure until only 150 years ago, a clock tick in the history of *Homo sapien's* time on Earth.

Modern humans no longer need to co-exist passively with our biological enemies, relying solely on the ability of the immune system or the acquisition of protective mutations to overcome infection. We have augmented our natural immune defenses by developing drugs that kill bacteria and viruses before they kill us, and have learned how to prevent the spread of infectious diseases by wearing condoms, washing our hands, and doing our best to separate our drinking water supply from our sewage. We have eradicated some major diseases by tricking the immune system with vaccines. Even as microorganisms evolve the means to bypass our antimicrobial artillery by acquiring genes that provide resistance to our most sophisticated arsenal of medicines, we have the ability to recognize the problem, and in most cases, to counter-attack. We have gained essential knowledge about some of infectious agents that have emerged recently in the modern era–HIV, Lassa Fever Virus, West Nile Virus, Ebola, and SARS– in record breaking time, which has allowed scientists and doctors to rapidly develop preventive measures and treatment. Even though AIDS has killed tens of millions of people, there are tens of millions of others free of disease who would have been infected if the underlying virus had not been quickly identified, and its mode of transmission established. Every year, tens of thousands of infants are saved from HIV disease by treating their infected mothers with antiretroviral medications before delivery. Development of new antibiotics has lagged a little in recent years, but it is more of an economic problem than a scientific one. The technology behind improved sanitation and advances in nutrition and food distribution, have also played major roles in controlling infectious disease.

The battle of the genomes that had provided infectious microorganisms with something of an upper hand for most of human history has shifted in our favor in the modern era, although microbes are certainly counterpunching with their genetic agility. After one hundred thousand years of fighting off a host of bacteria, fungi, protozoa, and viruses, *Homo sapiens* can only now claim an advantage, albeit a very provisional one.

The challenges now are to maintain our scientific edge and public-health vigilance, and to overcome the economic, political, and social forces that limit our ability to adequately share our gains with the world's poor nations. There, as I will begin to describe in the next chapter, infectious microbes capable of causing deadly infections of epidemic proportions still rule.

CHAPTER **3**

The Mosquito Plague

For some of the marines sent to Liberia in 2003 to help protect American citizens caught in the crossfire of a political uprising, the buzzing heard around their heads was not a passing bullet fired by the warring factions of a troubled African nation. It was the sound of a mosquito in search of blood. Instead of receiving purple hearts for sustaining battle injuries, 53 marines ended up in the infirmary with malaria. The marines, many of whom had served in the mostly mosquito-free Gulf region before their stint in sub-Saharan Africa, had no problem keeping their weapons and communications equipment in fine working order, but were complacent to the risk of malaria and failed to take their antimalarial medications properly. They were all cured, but three became critically ill.

It is likely, given the law of probability, that some of the marines sent to Liberia managed to avoid contracting malaria, or contracted a more mild case, because they, like hundreds of millions of other people in the world, harbor mutations in genes that provide a degree of protection against the disease. The common denominator linking these people is that they, and their descendants, live or once lived in the malaria belt, an expanse of planet Earth that covers the Equator and much of the Northern and Southern temperate zones, where malaria has been one of the major causes of death for the past ten thousand years. These mutations, which primarily occur in genes that influence the structure or function of red blood cells, arose by chance during human evolution and expanded by Darwinian natural selection because they confer a survival advantage against malaria, especially the variety caused by *Plasmodium falciparum*, perhaps the single most prolific killer of the human race. It is one of four species of one-celled protozoa that take up residence inside human red blood cells–the other three are *P. vivax*, *P. ovale*, and *P. malariae*. Infection with any one of these parasites is serious, causing debilitating fever, chills, and drenching sweats that can completely soak through clothes and bedding. *P. vivax*, *P. ovale* and *P. malariae* are not usually fatal, but *P. falciparum* is, killing its victims by inducing infected red blood cells to stick to small blood vessels in the brain, setting up a fatal inflammatory reaction known as cerebral malaria.

Plasmodia have survived for tens of millions of years by invading the red blood cells of various vertebrates, feeding on hemoglobin, the oxygen-carrying protein present inside red blood cells, timing their reproductive cycle to coincide with the feeding habits of their mosquito partners. Though related to each other, the different types of *Plasmodia* have evolved to adapt to the variety of vertebrates they infect, from birds to mice to monkeys, as well as the different mosquito hosts they need in order to complete their complex life cycle.

P. falciparum still ranks as one of the top three causes of fatal infectious diseases in the world, along with *Mycobacterium tuberculosis* and HIV. About 40 percent of the world's population is at risk of contracting malaria, mainly the billions of people who live in Africa, Southeast Asia and other tropical regions. About 500 million people are infected every year, with as many as two million deaths, about one every 30 seconds. Most of the fatalities are caused by *P. falciparum* infections in sub-Saharan Africa. About 90 percent of these cases occur in children under the age of five, contributing, along with infectious diarrhea and respiratory infections, to the continent's high rate of infant mortality–40 percent in parts of tropical Africa. Pregnant women are also more susceptible to fatal malaria infections. Malaria has killed nearly 14 percent of all Africans during the past century.

Malaria has been endemic in human and monkey populations for eons. We, and our great ape cousins, have similar kinds of red blood cells and hemoglobin that mosquitoes and *Plasmodia* seem to find appetizing. Although different species of *Plasmodium* have adapted to specific hosts, some that infect monkeys can also infect humans, and vice versa. The revolutionary change in human culture that began about ten thousand years ago–the transition from small hunter-gatherer groups to large communities with an agricultural base–contributed to the spread of malaria. Mosquitoes lay their eggs in motionless pools of water. New breeding grounds may have arisen when dense forests were cleared for agriculture, creating still pools of water in irrigation ditches and animal tracks. Human population expansion began to surge with improved agricultural technology, resulting in more densely populated communities and providing an ever-increasing pool of potential blood meals for the mosquito hosts of *Plasmodium*, various species of *Anopheles* mosquitoes.

The first reports of malaria date to the very beginnings of recorded history. Early Sumerian writings from 6000 BCE describe the illness. Cuneiform tablets found in Nineveh (now Iraq) describe malarial-type fevers that rampaged through ancient Mesopotamia. The Chinese text Nei Ching Su Wen, the oldest medical book in existence, written in 2700 BCE, described the classic patterns of fever that recur every three or four days in the different types of malaria, and attributed the disease to an imbalance in yin and yang. Malaria was introduced to Greece in 600 BCE by traders, soldiers, and slaves arriving from malarious regions of Asia and Africa, and within two hundred years it became the region's biggest health problem.

Hippocrates first noted the association between malaria and swampy areas. Greek physician and philosopher Galen (ca. 130-ca. 200), who served at the imperial court in Rome and influenced the practice of medicine for 1,500 years, attributed the disorder to an imbalance in the body's four humors (blood, phlegm, black bile, and yellow bile), caused by a dank effluvium or miasma (rotting organic matter) rising out of the marshes. The belief that a miasmic gas caused malaria and other diseases was the reigning medical paradigm for centuries. The Romans initially called this illness *atria cattiva* (Latin for spoiled air), which later became *malaria* (foul air). One of the most malarious areas in the world was the region of marshes south of Rome, an area that was uninhabitable until the 1930s, when Benito Mussolini, Italy's fascist dictator, had them drained.

Malaria was a serious health problem in the United States and North America until the 20th century. Europeans and African slaves brought the disease to the New World. The first known cases in the Western hemisphere followed Columbus' second voyage in 1493. He and most of his crew were stricken after being bitten by mosquitoes carried over from Europe, after which the malaria-causing protozoa were transmitted from infected crew members to local mosquitoes, and from there to indigenous populations. The early American settlement in Jamestown, Virginia, founded in 1607, was also hit hard by malaria, which killed as many as half of its residents and may have prompted the move to Williamsburg that occurred late in the 17th century. A Scottish visitor to the area in the early 1700s noted that "the fevers" were so bad, the region was good only for doctors and ministers.

Like the American marines in Liberia, fighting forces throughout history have had to contend with malaria. In 1798, 10,000 of Napoleon's troops died of malaria while serving in the eastern Mediterranean city of Acre. British troops have battled malaria throughout their nation's history of colonialism in Asia and Africa. In America, one of the first military expenditures passed by the Continental Congress during the Revolutionary War was $300 for George Washington to purchase the antimalarial drug quinine for his troops; the future president himself was stricken with an attack. During the Civil War, more than one million soldiers–nearly half of the Union and Confederate troops–suffered from malaria, killing about 8,000 of them. During World War II and the Vietnam War, malaria was a common cause of debilitating illness but only rarely caused death. More than one hundred thousand cases were reported in American troops during World War II, mainly among those stationed in Southeast Asia and the Pacific Islands (including Lieutenant John F. Kennedy), resulting in three million sick days and ninety deaths. General McArthur lamented, "This will be a long war if for every division I have facing the enemy, I must count a second division in the hospital with malaria and a third division convalescing from this debilitating disease." Nearly 25,000 military personnel were infected during the Vietnam War.

Explorers in Africa and other tropical regions also had to contend with malaria. Almost every one of them had one or two bouts; it was as much a part of the landscape as encounters with natives and exotic flora and fauna. A vial of quinine tablets became part of the traveling drugstore that explorers brought with them on their journeys. The ubiquitous gin and tonic (which contains quinine) provided a small amount of antimalarial protection, and an excuse for drinking to excess.

While malaria rampages through sub-Saharan Africa and parts of Southeast Asia, today the disease is a medical curiosity in the United States, now seen almost exclusively in Americans who travel to endemic areas and in foreign visitors to the United States. About 1,000 cases are reported every year. The urbanization and suburbanization of America helped rid the nation of malaria through swamp drainage, housing and highway construction, and the insecticide DDT, which destroyed mosquitoes and mosquito habitats, temporarily halted malaria's run as an endemic disease on American soil in the 1950s. Perhaps the single most important factor in the disappearance of malaria from the United States has been the widespread use of window screens that keep mosquitoes out of the home. However, since 1992, according to the Center for Disease Control (C.D.C.), there have been about two dozen cases of malaria in Americans who never traveled out of the country, including a couple of youngsters who live on Long Island, New York, near the largest urban and suburban centers in the United States. These cases represent infections that jumped from infected people, who acquired the infection overseas, into local mosquitoes. Several *Anopheles* mosquito species in the United States are quite capable of transmitting *Plasmodium*. One of the consequences of global warming in the 21st century will be an increase in mosquito habitats in the northern temperate zone, and it is possible that locally cultivated malaria will become more of a problem.

Because of its rarity, a case of malaria is an event for the modern American physician. As a medical student in the mid-1970s, a few cases were seen in servicemen returning from Vietnam who got infected either from mosquito bites or by injecting themselves with heroin using parasite-contaminated bloody needles, thus bypassing the mosquito vector exploited by *Plasmodium* during its tens of millions of years of existence. Even though we rarely saw a case, my young colleagues and I had learned a sufficient amount about malaria as medical students to suspect the illness in a Vietnam veteran who had just returned home and had developed high spiking fevers every three days. A few of us rushed off to the lab where students and interns examined blood, urine, sputum and spinal-fluid specimens, and were able to make a diagnosis of malaria by looking at a stained blood smear under the microscope before the clinical lab reported back to us. It was thrilling. The intracellular parasites were plainly visible in a few red blood cells, organisms that were stained a beautiful pale purple and easily distinguished, even to a novice, from some of the other purple-staining structures in a blood smear, such as platelets. They were *Plasmodium vivax*. It was hard to imagine that something so aesthetically pleasing to the eye could be responsible for so much suffering and death.

The excitement of finding a rare case of malaria on the wards of an inner city hospital, contrasted with my experience in West Africa. In the United States, the few cases the author treated were in adults, and all of them responded to antimalarial medication. However, in Africa, the diagnosis was more ominous. There, within the primitive hospitals that resembled relics from the Middle Ages, where relatives of patients were camped out on the grounds cooking meals out of giant cauldrons for their hospitalized family members, and where the cockroaches were so big you could hear the sound of their legs against the hard floors, the author saw small children infected with *Plasmodium falciparum.* They were so debilitated they laid hunched over their mother's shoulders like bags of flour, too weak to move or nurse, most destined for fatal cerebral malaria.

The idea that decaying swamp matter causes malaria persisted until the late 19th century, when the complicated life cycle of *Plasmodium* was unraveled through a series of brilliant natural and experimental observations. Two of the most important pioneers in this discovery were Charles Louis Alphonse Laveran (1845-1922), a Frenchman, and Ronald Ross (1857-1932), an Englishman, who were both army physicians. Laveran became interested in malaria when he was sent to Algeria after completing a tour of duty in the Franco-Prussian War in 1871. There, using a primitive microscope, he observed *P. falciparum* for the first time in a fresh sample of blood from a soldier with malaria.

Laveran's findings were dismissed by the French Academy of Medicine when he submitted a report, in part, because Laveran had no medical research credentials and his sketches were unconvincing (he was not a good artist). Drawings were important because the quality of most available microscopes was poor, and photographic microscopy had not yet been invented. The findings contradicted the view of two prominent scientists, Edwin Klebs and Corrado Tomasi-Crudeli, who claimed that they had identified the cause of malaria as a bacillus bacterium recovered from swamp water.

In 1884, German optician Carl Zeiss invented the oil-immersion lens, which increased the power range of microscopes, and a Russian chemist named Romanofsky discovered a stain that made it easier to view blood smears under the microscope. This allowed scientists to easily observe the parasite discovered by Laveran. Using the new staining procedure, Italian scientists were eventually able to visualize the four different types of *Plasmodia* that infect humans and thus confirm Laveran's observations, although, convinced of their superiority in all endeavors related to malaria, they often failed to acknowledge him. Laveran was finally recognized for his discoveries when he was awarded the 1907 Nobel Prize in Physiology or Medicine.

Initially, Laveran and others believed that humans acquired the parasite by drinking contaminated water. However, by the mid-1880s, he became convinced that mosquitoes transmitted the disease. Early travelers to Asia and Africa had been told by some natives that malaria was caused by mosquito bites, and some physicians were promoting this view. The idea that an insect or animal could carry parasitic diseases was given some impetus by Alexei P. Fedchenko, who in 1870 showed that the larval form of guinea worms is transmitted by freshwater crustaceans living in contaminated waters.[6]

The first mosquito-borne infectious disease ever described was elephantiasis, an affliction characterized by grotesque swelling of affected body parts. A prominent British parasitologist, Patrick Manson (1844-1922), showed that mosquitoes carried the microscopic filaria worms that cause elephantiasis, transmitting them to humans during a blood meal. These eventually grow into roundworms that obstruct lymph vessels, causing a backup of lymph and plasma, and massive swelling; the most commonly affected body parts are the legs and scrotum.[7]

Manson's discovery presented the possibility of a similar route of transmission for malaria, a cause taken up by Ross. Ross was an average student at St. Bartholomew's Hospital medical school, London, and an arrogant young man. He gave no indication of the experimental talent that would lead to one of the more important medical discoveries of the 19th century and a Nobel Prize. Ross painted and wrote verse, and viewed himself more as a poet than medical scientist. Because he had failed to distinguish himself in his medical studies, Ross chose to join the Indian Medical Service in 1881 to advance his career. In India, he became interested in malaria, and in 1892 began trying to find the disease's cause and mode of transmission.

Despite the findings of Laveran and the Italians, Ross initially believed that malaria was caused by an intestinal infection. However, he switched completely to the blood parasite idea when he was shown microscope slides of *Plasmodium*-infected blood, and ultimately became convinced that mosquitoes played a role. In 1895, Ross devised a clever experiment to prove this hypothesis, allowing mosquitoes to feed on malarial patients and then dissecting the insects serially over a few days to see if he could find living *Plasmodia*. His initial experiments failed. Due to his zoological ignorance, Ross was unable to distinguish the

---

[6]The larvae of guinea worms eventually turn into two-foot-long worms that creep throughout the human body and emerge through the skin.

[7]In some parts of the world, giant elephantiasis-swollen scrotums, which sometimes are so heavy they have to be carried in a wheelbarrow, commands a certain degree of respect and awe.

different mosquito species in the area. The common local mosquitoes, while abundant and easy to catch, did not transmit malaria. He spent two unsuccessful years dissecting the wrong breeds. In 1897, he was provided with an unusual mosquito he had never seen before, a member of the *Anopheles* family, and allowed the mosquito to feed on an infected Indian patient. A few days later, he dissected the insect's stomach and noticed a cystic structure filled to the brim with live *Plasmodia*. It was the first time the parasite had been observed outside an infected host's body. To study the problem more easily, Ross switched to avian malaria in sparrows as a model for the human disease.

Through Ross' findings, the idea of a mosquito vector for malaria began to circulate in tropical disease circles. However, some medical researchers believed that the illness was transmitted by swallowing infected water in which mosquitoes had dropped dead. However, Ross showed that active *Plasmodia* migrated from the cystic structure he had observed in the mosquito's gut to the insect's salivary gland. Female *Anopheles* mosquitoes have two hollow tubes in their proboscis, one for sucking blood and one for secreting saliva. The saliva contains enzymes and anticoagulants that prevent blood meals from clotting in the narrow proboscis, which would block the flow of blood and starve the insects to death. The saliva also contains anti-inflammatory agents that help anesthetize host skin to the blood-sucking mosquito, providing it with the one or two minutes it needs to feed without being swatted. Ross showed that infectious *Plasmodia* are spit into the host's bloodstream from the mosquito salivary gland during feeding.

In July 1897, an Italian researcher, Giovanni Batista Grassi, confirmed and extended Ross' findings, but he tried to take all the credit. Grassi, along with two other Italian malaria researchers, Amico Bignami, and Giuseppe Bastianelli, accomplished the very difficult task of defining the different developmental stages of malaria parasites; the protozoan can assume several forms in blood that are so distinct, they look like different organisms.

Grassi's arrogance got the better of him later on. Robert Koch had come to German East Africa to study malaria and stopped off in Italy to do some experiments. Grassi felt slighted when the Italian government provided the great Koch with resources to do the research. In one of Koch's rare failures, the experimental techniques he had developed to cultivate some of the most important bacterial causes of disease were ineffective for *Plasmodium*, which require living red blood cells for growth in the laboratory. Later, Grassi sent Koch a copy of his classic paper describing the role of *Anopheles* mosquitoes in human malaria. However, the tone of Grassi's letter offended the proud Koch. In 1902 when the Nobel Prize awards committee was leaning toward giving a shared prize to Ross and Grassi, Koch threw his considerable scientific weight against the Italian and the award was given, unfairly, only to Ross.

Through the work of Ross, Laveran, and Grassi, we now understand how malaria is transmitted and how *Plasmodia* mature in the red blood cells of mammals and the stomach of mosquitoes. A female *Anopheles* mosquito looking for a blood meal for nutritional support needed to lay her eggs, deposits thousands of immature forms of the parasite, called sporozoites, into host blood. The sporozoite first travels through the bloodstream and invades liver cells. Sporozoites mature into merozoites, which burst out of the liver cells after about two weeks, causing the first symptoms of malaria—high fever, headache, and teeth-chattering chills, followed by a drenching sweat. Within minutes, individual merozoites attach to and enter red blood cells, where they begin to feed on hemoglobin. Over the next few days, the merozoite matures into the trophozoite form, and most of these then divide into discrete merozoites, between 8-24 in number. However, a small number of trophozoites are programmed to develop into gametocytes, the male and female sexual forms of the organism. Infected red blood cells eventually burst open, which causes another round of symptoms, filling the blood with merozoites that subsequently invade other red blood cells, reinitiating the red blood cell phase of the life cycle.

To complete the life cycle, a feeding female *Anopheles* mosquito takes up male and female gametocytes, which fuse in the stomach to form a zygote. The zygote migrates through the gut and forms a cyst, easily observed with a low-power microscope, as first demonstrated by Ross. The cyst swells with thousands of infective sporozoites that burst out and migrate into the mosquito's body cavity and salivary gland. Sporozoites pass into the bloodstream of a new host when the *Anopheles* mosquito feeds again, starting the entire cycle all over again.

Blood meals consumed by a female *Anopheles* mosquito, during which she may increase in weight by several-fold, provide her with nutrients needed to mature and lay from 50 to 200 hundred boat-shaped eggs. The eggs are deposited in motionless pools of water, and float, like a baby with water wings, with the aid of air-filled sacs on their sides. These eggs mature into larvae in a few days and begin to consume bacteria and protozoa for nutrition by sieving water. Once the mosquito has matured, it is ready to mate and lay eggs. The entire life span of a mature *Anopheles* living in the tropics, during which it may infect several human targets with its parasite cargo, is only between 10 and 14 days.

Although it is not clear how mosquitoes and *Plasmodia* began to adapt to one another, it is likely that eons ago a mosquito larvae consumed a passing *Plasmodium* that had the capacity to resist destruction by the maturing insect's digestive enzymes. The protozoan that found itself trapped in the body of a developing mosquito had inadvertently stumbled upon its perfect partner, which shared its love of mammalian blood. What may have begun quite innocently as an indigestible microscopic meal in a tranquil tropical swamp evolved into one of the most relentless killers of Earth's most dominant species.

The aftermath of this ancient encounter is still observed today in the millions who die every year from malaria, and in the thousands who die from the common genetic disorders that the malarial infection spawned.

One of the distinguishing features of malaria is the regularity of the fever it produces, a finding so characteristic that it was noted in antiquity. It takes 72 hours for the merozoites of most plasmodial species to mature inside of red blood cells then burst out into the blood stream. The process is so synchronized that all the red blood cells swollen with merozoites–billions of cells–burst open simultaneously, producing the characteristic malaria symptoms of chills, fever and profuse sweating. The fevers occur like clockwork every three days in cases of falciparum, vivax, and ovale malaria (thus known as tertian malaria), and every four days in patients infected by *P. malariae* (quartan malaria); those infected by two different strains may experience daily fevers (quotidian malaria). In cases of tertian malaria, the attack begins at around noon every third day with shivering and headache, progressing to fever and sweating late in the evening. Between attacks of fever, malaria-stricken patients may feel normal, except for exhaustion. The remarkable synchronicity of this phase of parasite development is designed to coincide with both the feeding cycle of mosquitoes, which usually bite at night, and the maturation of *Plasmodium* gametocytes, which progress to the sexual phase of the life cycle 48 hours after the parasite enters the red blood cells, and which must be consumed in the blood meal in order for their sexual union to occur in the mosquito gut. So regular are the attacks of fever and chills that Sir Walter Raleigh, stricken with malaria and under a death sentence for treason, allegedly arranged for his execution to take place on a day he knew would be fever-free, to avoid the appearance of cowardice should his teeth start to chatter from an attack of malarial fever and chills.

Depending on the victim's immune status against malaria, attacks of fever may occur once or twice before the infection is under control, or it may flare up a dozen times or more.

An interesting aside to the malaria story is how the high fever caused by *Plasmodium* infection was exploited to control neurosyphilis, an end-stage complication of untreated syphilis. Julius Wagner-Jauregg (1857-1940), a Viennese internist and psychiatrist, developed an interest in treating mental disorders by inducing fever after he observed that some mentally ill patients seemed to improve after acquiring infections. After unsuccessful attempts at inducing fever using tuberculin (a protein extract from the tuberculosis bacillus), he tried a different approach on patients suffering from neurosyphilis, infecting them with *P. vivax*, which he knew could be controlled with quinine. It worked: malaria therapy arrested the neurosyphilis. Later, the treatment was altered by replacing *P. vivax* with *P. knowlesi* (which causes malaria in monkeys), because it produces a self-limiting infection in humans. For 25 years, malaria therapy was an important treatment for controlling neurosyphilis, and it earned Wagner-Jauregg a Nobel Prize in 1927. The treatment was made obsolete with the advent of penicillin in the 1940s, which penetrates into the brain and effectively kills the bacterium that causes syphilis, preventing or arresting neurological complications.

Although Wagner-Jauregg's name has faded into history, malaria therapy, together with the observation that epileptic seizures improved symptoms of depression, provided part of the intellectual framework that led to the development of insulin shock therapy–and ultimately electroconvulsive shock therapy–as a method for treating refractory depression.

The historic treatment of malaria is quinine, a natural alkaloid found at high concentrations in the bark of the cinchona (pronounced *sin-cone-a*) tree, which is indigenous to South America's Andes Mountains. Its medicinal properties were introduced to Europe in the 1630s by Spanish Jesuit missionaries. The priests had learned that the Indians stripped the bark off the tree (which they called quina-quina) to use as a medicine. According to a popular historical legend, an extract of the bark was used to successfully treat the wife of the Viceroy of Peru, the Countess of Cinchon, who was suffering from malaria (a story debunked in 1941 by an American scholar, who claimed that the countess never had malaria and died of another illness during her voyage back to Spain)[8]. Carolus Linnaeus, the great Swedish botanist and physician who in the mid-18th century, developed the genus/species system of classifying plants and animals, immortalized her name by Europeanizing quina-quina into *Cinchona* as the name for the tree's genus.

The Jesuits brought the bark extract to Rome and the Vatican where it was called Jesuit's Powder. Sensing its importance as a medicine, the Jesuits instructed the Indians to plant five cinchona trees for each one that was harvested. Conservation was not the only motive; the Indians were instructed to plant the five trees in the configuration of a cross.

---

[8]Norman Taylor, Cinchona in Java, p. 30 (New York: Greenberg, 1945)—from www.chemweb.calpoly.edu/chem/bailey

There was Protestant resistance to cinchona bark in Europe, where it was viewed as a papal plot. There was also resistance by the medical establishment because its use conflicted with the ideas of treatment espoused by Galen who recommended bleeding and purging to restore "balance to the humors." The Galen remedy was undoubtedly fatal to many malaria sufferers: bleeding worsened the anemia typically found in malaria (anemia is caused by the relentless destruction of red blood cells), and purging worsened the dehydration caused by the fever and sweating. The poor, who could not afford to see a physician to get the Galen treatment, had a better chance of surviving an attack.

Resistance to cinchona may have played a minor role in British history. Oliver Cromwell, leader of the forces against the British monarchy during the English civil war (1642-48) and England's lord protector (1653-58), is believed to have died of malaria, stricken with "tertian ague" that would have likely responded to quinine had it been available. After Cromwell's death in 1658, resistance against the deposed monarchy crumbled, and in 1660, the Stuart line assumed power in England under Charles II, son of Charles I (who had been tried for treason and beheaded by Cromwell's followers in 1649).

Charles II helped break the anti-Catholic prejudice against cinchona and popularize its use throughout Europe after he was cured of malaria by a concoction brewed by the medical quack Robert Talbor. The entrepreneurial Talbor combined cinchona bark extract with wine to mask its bitter taste, and opium for its euphoric and analgesic properties. Charles II was so impressed with the treatment that he appointed Talbor as court physician, despite his lack of an official medical education. Later, Talbor visited France and cured the son of Louis XIV of malaria. After Talbor's death, his formula was revealed and cinchona bark became a staple of malaria treatment for those who could afford it.

The commercial use of quinine expanded because of the entrepreneurship of Charles Ledger (1818-1905) an Englishman who was sent to the Andes in 1836 to explore the commercial possibilities of alpaca wool. Ledger became intrigued with cinchona and finally found one particular variety with extremely high quinine concentrations. With the assistance a Bolivian native named Manuel Incra Mamani, an expert on cinchona who helped him collect bark and seeds between 1844 and 1865, Ledger smuggled 14 pounds of seeds of that species (which was subsequently named *C. ledgeriana*) to Europe in 1865. The British government was not interested, but the Dutch, whose colonizing efforts in Asia were being thwarted by high rates of malaria there, bought one pound of the seeds from Ledger for 100 guilders (about 20 dollars) and planted them in Java. The cinchona trees that sprang from these seeds soon provided the Dutch with a worldwide monopoly in the supply of highly concentrated quinine, effectively destroying the market for South American cinchona. Mamani was jailed, beaten, and eventually starved to death for his treachery. Javanese cinchona helped the Dutch in their colonization efforts and continued to be the major source of quinine for nearly a century, until Japan's occupation of Java during World War II. Cinchona is now grown in Africa and Southeast Asia, and is still cultivated in the Andes.

Despite its long history, quinine is not an ideal antimalarial. Although it has saved many lives, it does not cure malaria in everyone, it is more effective as a prophylactic agent, and it can have toxic side effects at doses needed to kill malaria-causing *Plasmodia.* It can also cause disturbances in vision and balance. Quinine was commonly used by drug dealers to dilute or "cut" heroin and has also been implicated in toxic reactions experienced by intravenous drug users, some of them fatal.

The German army's numerous malaria casualties in World War I, despite the use of quinine by military personnel, proved to be a major impetus for the development of new and more effective antimalarial drugs. During the 1920s, many such compounds were synthesized and tested such as plasmoquine, the forerunner of the antimalarial drug primaquine, and sontochin, a precursor of chloroquine, which was developed by Americans at the end of World War II, and which subsequently became the most important antimalarial drug in the world. World War II provided incentives for the United States and other European nations to find new antimalarials when the major source of quinine was cut off after the Japanese occupied Java and the Germans confiscated millions of pounds of quinine stored in a facility in Amsterdam after they occupied the Netherlands.

Malaria was also a problem for Germany's troops in Africa.

There is no small irony in the fact, discovered over the past decade by genetics researchers, that the unique genetic variants that provide survival-of-the-fittest protection against *Plasmodium falciparum*, which will be described in the next few chapters, are almost exclusively found in non-Germanic people, a fact that would have shocked Nazi proponents of Aryans as the master race.

The search for more effective malaria treatments took on a cruel twist during World War II. In addition to performing barbaric and bizarre human experiments on the effects of cold temperatures and high altitudes on the body, and testing different sterilization methods, using prisoners in concentration camps as subjects, the Nazis tested toxic concentrations of antimalarial compounds combined with other drugs. The site chosen for this assignment was the concentration camp in Dachau, Germany. Karl Schilling, a former director of the Robert Koch Institute for Tropical Medicine was selected by SS Leader Heinrich Himmler to lead the project. To test the efficacy of Schilling's treatments, about 1,200 inmates were either injected with malaria-contaminated blood or exposed to malarious mosquitoes by attaching boxes of the hungry insects to their arms. Many of the concentration camp prisoners who were forced to participate in these experiments died from the toxic side effects of the ill-conceived drugs. After the war, Schilling was hanged for crimes against humanity.

American medical and military personnel obtained sontochin from Vichy French physicians after Tunis was liberated from the Germans in 1943, and derivatives of the drug were developed, the most promising of which was chloroquine. Chloroquine satisfied all the requirements of a wonder drug. It was very effective, it produced no significant side effects, and it was inexpensive (about 8 cents a treatment). Within a short time, chloroquine became the treatment of choice. Its widespread use beginning just after World War II resulted in a significant decline in mortality from malaria throughout the world and, coupled with the use of insecticides to kill mosquitoes, humans, for the first time in history, gained control over its most pernicious infectious disease enemy.

In the late 1950s and early 1960s, however, resistance to chloroquine began to emerge in *Plasmodia* strains found in pockets of South America, Southeast Asia, and Africa, and slowly spread around the world. Resistance to chloroquine progressed by Darwinian selection through a series of mutations in a *Plasmodial* gene that provides the parasite with a way of degrading the drug, a chemical scissor specific for chloroquine. Now, chloroquine resistance is so widespread that it is useless in the majority of cases.

Becoming resistant to its most potent chemical enemy is just one of many genetic tricks in the *Plasmodium* arsenal. One is based on the very core of the *Plasmodial* life cycle; its ability to assume different physical forms in an infected host, allowing it to evade destruction by creating a moving target for the immune system. An immune response generated against sporozoites, for example, may be ineffective against merozoites because of differences in the proteins expressed during these different phases.

Another trick involves a sophisticated interaction with the infected host's immune system utilizing a protein called P-fEMP (plasmodium falciparum erythrocyte membrane protein), which also happens to be one of the key factors responsible for the development of cerebral malaria. P-fEMP is transported to the surface of infected red blood cells, which induces them to attach to blood vessels. When this happens in the brain, an inflammatory response is generated which can cause irreversible damage. Forcing red blood cells to attach to blood vessels is advantageous to the parasite because it reduces the likelihood that its red blood cell host will be consumed by phagocytic cells located in the spleen, an organ of the immune system that doubles as a filter for aging, damaged, and infected red blood cells. Sacrificing infected cells is one way the immune system eliminates intracellular infectious microbes. By attaching to blood vessels, infected red blood cells spend less time in the circulation, and therefore make fewer transits through the spleen, effectively avoiding the phagocytic cells that lie in wait there. On the other hand, by forcing red blood cells to transport P-fEMP to the surface, the parasite exposes infected cells to an attack by antibodies and T-lymphocytes.

It seems like a reasonable trade-off for the parasite: the same protein that exposes it to an immune system attack also helps it to avoid being destroyed by phagocytic cells in the spleen. However, not satisfied with a mere trade-off, *Plasmodium* has evolved a strategy for countering the immune attack provoked by P-fEMP. There are 60 copies of the gene that codes for P-fEMP in the parasite's genome, instead of one or two-the number of genes expected from the different stages of the parasite, which have either one or two copies of every chromosome. Each one produces a slightly modified version of P-fEMP protein capable of performing its critical function in the parasite's life cycle; i.e., attachment to the lining of blood vessels to avoid destruction in the spleen. However, they differ sufficiently in their amino acid sequence so that the antibodies produced against one variant will not react against another. Like the many heads of the mythical hydra, for every parasite that is knocked out because it expresses a particular P-fEMP variant recognized by antibodies, others take its place.

The variability of P-fEMP is one reason for the difficulties that researchers have encountered in developing an effective malaria vaccine. For a vaccine to work, it must generate an immune response capable of handling the different P-fEMP variants expressed in *Plasmodium*.

Eventually, if a person infected with *Plasmodium* survives the initial attack, the immune system can take control once the parasite has exhausted all its genetic tricks. Malaria immunity, when it does finally develop, is not perfect. Another attack is possible, although it tends to be relatively mild. There is no cross immunity between different *Plasmodium* strains; thus, if more than one strain is endemic in an area, one infection can immediately follow another, or both can infect the same individual simultaneously. Although imperfect, natural immunity does play an important role in keeping malaria in check. Populations with low levels of malaria who have little natural immunity can be devastated by the sudden introduction of a large population of *plasmodium*-filled mosquitoes.

The mosquito carriers of *Plasmodia* are also highly adaptable, which has led to the spread of malaria all over the world in a variety of different animals. There are 3,200 different types of mosquitoes on Earth, but only *Anopheles* carries the parasite that causes human malaria. Of the 430 different *Anopheles* species, about 70 are capable of transmitting *Plasmodia*. *Anopheles* adapt to a variety of different environments, but are not found in mountainous regions 2,500 meters above sea level, where the temperature is too cold for this species to survive. *Anopheles* are delicate in this regard, in contrast with the various mosquito species that are capable of living in more inhospitable regions, such as the Arctic Circle, home to some of the largest mosquitoes in the world. Feasting hordes of these cold-loving insects, which bite ravenously during the few weeks of the year when the air temperature rises enough for the frozen tundra to thaw, can suck nearly the entire blood supply out of a large caribou.

Anopheles is particularly attracted to humans and easily adapt to the habitats we artificially create for them. Any warm, still pool of water can serve as a nesting ground for mosquitoes to lay their eggs. Foot prints, tire tracks, open beer cans, and the inner hollow of old tires can all fill with rain water and be warmed by the sun to create makeshift homes for these eggs to turn into developing larvae. Mosquitoes thrive in these makeshift habitats, which are free of the predatory fish that inhabit lakes, rivers, and streams and feed on larvae. In addition, adult mosquitoes can be transported for miles from their natural habitat by wind; and sometimes they hitch a ride with travelers—a few people waiting in airports have been stricken with "airport malaria" after being bitten by mosquitoes brought by planes that departed from malaria endemic areas. In the summer of 2003, the C.D.C. reported an outbreak of seven cases in Palm Beach County, Florida, all of them living within 10 miles of the Palm Beach International Airport and infected with the same strain of *Plasmodium vivax*, according to genetic analysis.

Both extreme rain and drought can produce new mosquito-breeding habitats. Heavy rains, which create freestanding puddles perfect for egg laying, may have resulted in 150,000 malaria deaths in Ethiopia in 1958. Severe droughts dry up running rivers and leave still pools of water that can support the fragile eggs, and reduce the population of mosquito larva predators. In 1934-35, a drought in Sri Lanka (then called Ceylon) may have been responsible for 80,000 malaria deaths. The El Niño weather system, a warm, dry mass of air originating in the Western Hemisphere, has caused changes in temperature and rain in Africa that has been a boon for *Anopheles* and *Plasmodium*. El Niño has become more severe over the past decade, probably because of global warming.

Traditionally, destroying *Anopheles* breeding grounds has been an effective means of reducing mosquito populations. W. C. Gorgas drained swamps in Panama and thus reduced yellow fever and malaria, allowing workers to complete the Panama Canal project. The British drained swamps in Malaya in order to help destroy mosquitoes that were preventing them from establishing rubber plantations. However, draining swamps is expensive, and economic incentives are usually needed to drive these projects. There has been no incentive to destroy mosquito-breeding grounds in sub-Saharan Africa.

The world needs other inexpensive and efficient ways to destroy *Anopheles* mosquitoes. For many years, the insecticide DDT seemed to be the best solution to this problem.

It was initially synthesized in 1874 by an obscure Swiss pharmacist named Othmar Ziedler. The formula was rediscovered in 1939 by Paul Müller, a Swiss chemist working at Geigy Pharmaceutical, who was searching for an effective and inexpensive insecticide that would kill, of all things, clothes moths. DDT fit the bill. It was the most effective insecticide ever discovered; a potent insect neurotoxin. It is also extremely stable– a single spraying is capable of protecting a house from mosquitoes for several months, at doses that are not toxic to humans. However, its chemical stability ultimately contributed to its downfall. In a pilot project in the United States, DDT was used to successfully eradicate malaria from the Tennessee River region, one of the last regions in the United States where the disease remained endemic into the 20th century. The success of this and other projects led to the Global Eradication of Malaria Plan sponsored by the World Health Organization (WHO), and garnered a Nobel Prize for Müller in 1948, so great was the promise of DDT.

Widespread use of DDT resulted in a significant decline in malaria around the world, as well as to a decline in other insect-transmitted diseases, such as Kala Azar (visceral leishmaniasis), a chronic and often fatal parasitic infection transmitted by sandflies, and typhus, a bacterial infection spread by lice, the bane of soldiers, refugees in settlements, and prisoners of war. The common use of DDT as a non-specific insecticide, killing everything from moths to mosquitoes, had some unintended consequences, including the destruction of the roofs of huts in Malaya, as described by Robert Desowitz in *The Malaria Capers*. When a malaria-eradication program was initiated in one area there, roofs made of attap, a large leaf from a Malayan palm tree, began to collapse because the DDT destroyed a parasitic wasp that preyed on a caterpillar that fed on attap. The unintentional elimination of the wasps led to an overgrowth of the caterpillars, which voraciously consumed the attap roofs.

DDT's effectiveness in fighting malaria in some parts of the world was very dramatic. In India, the number of cases during the 1950s and 1960s fell from 75 million a year, resulting in 800,000 deaths, to 100,000 cases annually, and a negligible number of deaths. Progress was also made in Africa. However, the initial success with DDT was short-ived. The Herculean task of attempting to kill countless numbers of mosquitoes living in the world's malaria-endemic areas became even more difficult when *Anopheles* began to fight back with its own survival strategy: DDT resistance. Resistance took two forms, behavioral and physiological. *Anopheles* mosquitoes rest on walls after feeding on humans. However, homes sprayed with DDT killed *Anopheles* that acted this way and selected for a behavioral variant that retreated to the outside after feeding, improving their survival. Physiological resistance emerged in *Anopheles* mosquitoes with a gene variant that increased their ability to destroy DDT enabling them to survive the insecticide assault.

The biggest blow to using DDT in the campaign to eradicate malaria was an emerging awareness of the insecticide's environmental impact, popularized in Rachel Carson's book *Silent*

Spring, published in 1962. DDT was implicated in the loss of songbirds and fish, and its potential to travel up the food chain alarmed environmentalists and the public. So widespread was the use of DDT in the 1960s–more a result of its use as a crop pesticide than for mosquito eradication–that trace amounts could be found in nearly every living human being on Earth. DDT was also found in fish and livestock, and went through the food chain, accumulating in animals, such as polar bears, that never came in direct contact with the insecticide. DDT is very stable and accumulates in body fat; its half-life in humans (the time it takes for the body to metabolize half an acquired load) is, alarmingly, eight years. Within a few years of the book's publication, a worldwide ban on the insecticide was put into effect.

The DDT ban may have been sound environmental policy, but it came at a price. Although DDT resistance was becoming a serious problem, it was still effective as an insecticide in most parts of the world. After the ban, the worldwide rate of malaria began to rise again. In India, millions of cases are again being reported every year. In sub-Saharan Africa, the DDT ban probably cost millions of lives; falciparum malaria reemerged, back to its historic deadly level.

To limit environmental exposure in combating malaria, insecticide-impregnated bednets instead of widespread spraying can be used. Mosquito nets work very well, but they are more expensive than spraying (the price is higher than necessary because some African governments have placed a tax on their purchase in order to increase revenue). Fewer than 2 percent of Africans currently use them. DDT is perhaps the best insecticide to use in bednets, but because of the ban, chemical companies are no longer synthesizing it and stocks are very limited. The most commonly used insecticide in bednets is pyrethroid. However, *Anopheles gambiae* is already becoming resistant it.

New strategies for fighting malaria will undoubtedly be devised from the wealth of information obtained from the recent genome sequencing of both *Anopheles gambiae,* the major mosquito carrier of human malaria, and *Plasmodium falciparum,* which will be discussed in the last chapter. There is even research on the feeding behavior of *Anopheles*, focusing especially on gaining an understanding of why these mosquitoes find some people more attractive than others; they have a penchant for smelly feet and are attracted to the odor of Limburger cheese, for example. There is some evidence that gametocytes may signal *Anopheles* by affecting the body odor or breath of infected patients, thereby attracting hungry mosquitoes to feed on people carrying the parasite forms needed to generate new progeny. Understanding the chemistry and biology of mosquito attraction may lead to better repellants and traps.

Vaccine research, which was stagnant until a few years ago, because of perceived low profitability, is now a very active area of research, thanks to the efforts of the Malaria Vaccine Initiative, and the Roll Back Malaria initiative sponsored by the United Nations and WHO. More than two-dozen candidate vaccines are in various stages of the clinical trial pipeline. Much of funding in this area has come from the Bill and Melinda Gates Foundation. But an effective vaccine is proving difficult to develop for the same reasons *Plasmodium* is able to escape the immune system–there are too many target proteins that need to be attacked.

Another non-profit organization, the Medicines for Malaria Venture, has been successful in getting some pharmaceutical companies, such as Novartis Pharma AG and GlaxoSmithKline, to invest in malaria treatments. The lack of investment in malaria research by major pharmaceutical companies in the 1980s and 1990s at the same time that chloroquine resistance was emerging, helped create the current flare-up of malaria. One important "new" drug is artemisinin, a potent antimalarial chemical derived from the Chinese herb *Artemesia annua*, which has been used as a fever remedy in China for more than 1,000 years. It is still used today in Southeast Asia as a traditional medicine. However, its high cost, short shelf life, and the lack of modern day clinical trial data have limited use elsewhere. Drug companies have developed chemical derivatives of artemisinin that are now used in combination with other antimalarials. These are rapidly becoming the drugs of choice for malaria treatment around the world. The success of artemisinin derivatives has spawned an illegal underground industry in Southeast Asia that has flooded local pharmacies with counterfeit drugs. The drugs are packaged to mimic products manufactured by legitimate pharmaceutical companies, but are either diluted antimalarials or, in the more blatant counterfeits, inactive placebos. In surveys conducted in 2003 throughout Vietnam, Cambodia, Thailand, Laos and Myanmar, more than half of the artemisinin-type drugs tested were fake. Organized crime gangs are behind it, and their tactics of using corruption money to bribe officials, police and pharmacists, and using the threat of violence to intimidate activists, suggest that the practice will not end quickly.

It is unlikely that any single method–vaccine, insecticide, or antimalarial drug–will be completely effective when used on its own. The ability of *Plasmodia* and *Anopheles* to reconfigure as resistant organisms suggests that combining different approaches offers the best chance for eradication. As for the role of insecticides in eradicating malaria, the perceived and real risks to the environment will have to be weighed against the loss of human life from a disease that kills one person every 30 seconds of every day.

CHAPTER 4

Killer Beans

The story behind one of the most common malaria-protective mutations in the world begins with a plant.

Long before there were genetically modified foods by Monsanto, nature created *Vicia faba*, the fava bean (also known as the broad bean), a nutritious dietary staple consumed throughout the Mediterranean region and Middle East. Fava beans are valued by nutritionists and vegetarians for its high protein content, the highest among the legumes after soy. They thrive in semi-arid land, do not require irrigation, and are extremely hardy, growing well enough in cool weather to provide a late winter-early spring harvest. The fresh bean, uncooked, is sweet and tender. The pod of *Vicia faba* is used in alternative medicine as a topical treatment for fungal infections of the skin; it also contains detectable levels of L-dopa, which, in a pure pill form synthesized by the pharmaceutical branch of DuPont, is the principal drug used to treat Parkinson's disease.

Unfortunately, in addition to the benefits provided by this dietary staple and medicinal plant over the centuries, it has caused severe illness in millions of people, and killed hundreds every year. Favism is the term used to describe the sensitivity that some individuals have to the beans and pollen of *Vicia faba*. This condition is found primarily in the Mediterranean region–among Italians, Greeks, Turks, Cypriots, inhabitants of the islands of Sardinia and Sicily, and in the Middle East, Southeast Asia,

Taiwan, and China. On the island of Sardinia, which has the highest prevalence of favism in the world, 250 deaths were reported in 1936. Before the sensitivity to this bean was discovered and preventive measures and treatment became available, attacks of severe lethargy in schoolchildren (primarily boys) were observed every February in Sardinia, brought on by the earliest harvest of fava beans. Locals attributed the problem to the "natural" laziness of the boys who were affected, and accused them of feigning tiredness to avoid schoolwork and chores.

Raw or undercooked fava beans and *Vicia faba* pollen are the main culprits. The toxic product can also be passed to susceptible nursing infants. The symptoms of favism–all due to the massive destruction of red blood cells–are severe fatigue and jaundice. Symptoms begin approximately 12 to 48 hours after exposure to the bean or pollen, striking without warning. This sensitivity tends to occur in families, especially in males. In the past, an attack of favism had a mortality rate approaching 10 percent. However, since transfused blood is resistant to the effects of fava beans (unless the donor also has the problem), the availability of safe transfusions since the early 20th century has markedly reduced the incidence of severe favism, and deaths should never occur. Public awareness of the condition is high in parts of the world where favism is common. During the harvest season, restaurants in Italy that serve dishes made with fava beans warn customers when the beans are in the kitchen. It is not good business to expose your male customers to them if 10 percent are likely to destroy their red blood cells after inhaling pollen drifting in from the vegetable bin or after consuming the fava bean-based house specialty. Recipes containing fava beans sometimes come with warnings, instructing readers that people of Mediterranean descent may have to consult a doctor before trying them out.

Sensitivity to fava beans has been around since antiquity. It has even been suggested that Pythagoras, the Greek mystic, musician, mathematician, and philosopher, suffered from it. Although his brotherhood was dedicated to vegetarianism, eating beans was forbidden. Most likely though, Pythagoras developed this aversion from his travels throughout Asia Minor, where eating beans was discouraged by some religious sects because of certain sexual connotations attributed to them. The philosopher Zarates, for example, discouraged men from eating beans because the smell of chewed, sun-cured beans was thought to resemble that of semen.

The connection between fava bean sensitivity and the destruction of red blood cells was discovered in the 1920s. However, it took another 30 years before the exact biochemical basis for favism was discovered, and another 30 years before scientists found the genetic cause behind it. The solution to the problem began when medical researchers were trying to understand another, seemingly unrelated, enigma; 10 percent of African-American men were developing jaundice and anemia after taking antimalarial drugs, especially primiquine. Black women were also affected, but not as severely as men. Rare cases of sensitivity were also reported in other ethnic groups. The symptoms of primiquine sensitivity resembled favism, only not as severe, causing a partial destruction of red blood cells. After a few days of illness, the symptoms resolved on their own, with complete recovery in almost all cases.

The problem of antimalarial sensitivity exploded during World War II because these drugs were being taken by troops stationed in malarious zones in the Pacific theater. A racial disparity in the sensitivity to antimalarial medications was also observed in troops from other countries. Among British and colonial soldiers, anemia and jaundice were seen in the Indian and Burmese troops, but not in the English. This difference was attributed to the smaller body size of colonial soldiers and to ethnic differences in the metabolism of antimalarial drugs.

In 1953, a simple experiment showed that sensitivity to primiquine was caused by an intrinsic defect in the red blood cells of susceptible individuals, and not by racial differences in body habitus or drug metabolism. In the experiment, red blood cells from individuals sensitive to primiquine were tagged with a very small amount of radioactive chromium (similar to the amount of radiation one would receive today when getting a nuclear scan), after which the tagged cells were given to non-sensitive recipients. Then, the recipients were given the offending drug. Their blood was shown to contain free radioactive chromium that was released from the tagged cells because they were being destroyed by primiquine, indicating that the donor red blood cells were the source of sensitivity.

In the 1950s and 1960s, scientists discovered that primaquine-sensitive red blood cells were deficient in their ability to handle chemical damage caused by oxygen. Red blood cells are packed with the oxygen-binding protein hemoglobin, which latches onto oxygen in the lungs and delivers it to every cell in the body. Oxygen is needed for the biochemical reactions used to generate large quantities of ATP (adenosine triphosphate), the major energy-storage molecule of life. However, oxygen can also damage cells by chemically oxidizing proteins and lipids. The rusting of iron is an example of an oxidation reaction; and the effects of aging on skin and other parts of the body are due, in part, to the damaging effects of oxidation on cells and connective tissue.

To counteract oxidation damage in red blood cells, various antioxidant systems have evolved, one of which uses an enzyme called glucose 6-phosphate dehydrogenase (G6PD), which is needed to generate the strong antioxidant molecule, NADPH. As long as G6PD is operating normally, a steady supply of NADPH is available to protect red blood cells against the damaging effects of oxygen. However, the antioxidant pathway only operates for a finite period of time–three to four months in red blood cells, after which the damaging effects of oxygen cannot be corrected and the cells die (the average life span of a red blood cell is about 120 days).

The G6PD antioxidant pathway was found to be deficient in people sensitive to fava beans and antimalarial medications.

Broken-down, oxygen damaged red blood cells are continuously being removed from the circulation, with new ones taking their place. In an individual who receives proper nutrition and an abundant supply of iron, folic acid, vitamin B6, and vitamin B12, and who has intact kidneys capable of producing the hormone erythropoietin (which is needed to stimulate the growth of red blood cell precursors in the bone marrow), replacement of dead red cells with fresh ones keeps the number of red blood cells stable over time. In fact, every day of our lives, nearly one percent of our oldest, oxygen-damaged red blood cells are removed from the circulation–tens of billions of cells–and replaced by an equal number of new, freshly minted ones.

Why does the antioxidant system in red blood cells become depleted after only a few months, contributing to the daily destruction of such a large number of cells, and why does exposure to fava beans and antimalarial drugs further accelerate the process in sensitive individuals? Although the life span of cells is determined by a number of different variables, the principal factor responsible for the finite life span of red blood cells is their unique intracellular anatomy. In these cells' first few days of life, intracellular organelles, the specialized structures that perform vital chores, such as mitochondria, which produce ATP, ribosomes, where protein synthesis takes place, and the nucleus, which contains all of an organism's genes (except for a few dozen found in mitochondria), are all eliminated from the maturing red blood cell.

With the nucleus and other organelles present for only the first few days of its life in the bone marrow, there is a period of very intensive gene activity and protein synthesis, during which all of the hemoglobin, antioxidant compounds, including G6PD, structural proteins needed to create the scaffold that maintains the cell's shape, and other proteins essential to sustain the red blood cell and keep it functioning for 120 days are synthesized. After this period of frenzied activity, the nucleus begins to degenerate; all gene activity stops. After another day, the nucleus and most of the organelles are wrapped with a piece of cell membrane and eliminated from the cell, their work done. Without a nucleus, no additional mRNA can be synthesized, and, without mRNA, no additional protein synthesis can occur. Thus, every protein found in a mature red blood cell was synthesized during its first few days of life. Without the capacity to refurbish itself by activating genes and synthesizing new proteins, the level of G6PD and other proteins decreases as the cell ages. As G6PD levels decrease, the red blood cell can no longer protect itself from oxidative damage; they wear out and die. This process is markedly accelerated when sensitive people are exposed to fava beans and antimalarial drugs, which cull the oldest red blood cells; those with the lowest level of G6PD.

The evolutionary driving force behind the "self-destructive" elimination of the nucleus and other organelles in mammals is enhanced delivery of oxygen to the rest of the body. Without organelles, red blood cells are more flexible and deformable, allowing them to squeeze through tiny blood vessels (capillaries) smaller than the diameter of the red blood cell. Close contact maximizes the surface area of interaction between the red blood cell's membrane and capillary walls, thus facilitating the transfer of oxygen. Also, a large nucleus physically interferes with the ability to accumulate maximal amounts of hemoglobin. As the hemoglobin level rises, so does the capacity to carry oxygen, improving an individual's endurance. This is why competitive athletes improve their performance through "blood doping," a method for artificially increasing hemoglobin levels by transfusing athletes with blood prior to an important event. In recent years, a number of long-distance runners, bicyclists, and cross-country skiers have been accused of blood doping when their performance level dramatically improved during the Olympics or other high-profile event. The same effect as blood doping can be achieved legally by athletes who train at very high altitudes, which increases the amount of erythropoietin in the blood naturally in response to the low oxygen level, boosting the production of red blood cells and raising hemoglobin levels. For the biotech-savvy modern athlete, an alternative to blood doping and the inconvenience and discomfort of high-altitude training is to inject themselves with bioengineered erythropoietin, which is used by doctors to treat the anemia of kidney failure, cancer, and AIDS. However, erythropoietin is a banned substance in competitive sports. To screen for blood doping and illicit use of erythropoietin, the governing bodies that oversee endurance events, such as the Tour de France bike race, set an upper limit for the allowable level of blood hemoglobin.

Natural selection has also exploited the advantages of increased hemoglobin levels, promoting the drastic step of eliminating organelles in mammalian red blood cells to maximize the accumulation of hemoglobin and the cells' oxygen-carrying capacity. In terms of evolutionary drive, the nucleus in the mature red blood cell–the cell's brain and control center–is just a bloated waste of space. The trade-off is the inability of mature red blood cells to synthesize new proteins when they undergo biological stress, which has harmful consequences for people with favism.

With the emergence of recombinant DNA technology in the 1970s, enabling researchers to isolate genes and determine their complete DNA sequence, the breakdown in the red blood cell antioxidant pathway that occurs in favism was found to be caused by inherited mutations in the gene that codes for G6PD, resulting in a reduction in G6PD enzyme activity. Over the past few years, we have come to realize that this deficiency–the strange condition that causes fava beans and certain medications to destroy the red blood cells in Africans, Italians, Greeks and people from the Middle East and southeast Asia–is one of the most common genetic abnormalities in the world, affecting hundreds of millions of people, as many as 5 percent of the human population.

There are several hundred different types of G6PD gene mutations that reduce the amount of G6PD enzyme available to red blood cells, and they essentially fall into three major categories: severe, moderate, and mild. In the severe form, which is quite rare, the amount of G6PD activity is so low that the ability to generate the antioxidant NADPH is compromised all the time, resulting in chronic destruction of red blood cells that keeps patients constantly anemic. In the moderate form, which causes favism, and which is common in Italy and Greece, there is a sufficient amount of G6PD activity so that the antioxidant pathway, while slightly impaired, can still function under normal conditions for the entire life of the red blood cell; these people destroy their red blood cells only after experiencing an oxidative stress crisis, which occurs after they consume fava beans, which, it turns out, contain several oxidizing chemicals. Antimalarials and other drugs that have oxidizing properties will also destroy red blood cells in affected individuals.

In the mildest type of G6PD deficiency, which is found in 10 percent of African-American men, the oxidative stress condition that causes the destruction of red blood cells occurs only when very strong chemical oxidizing agents are consumed, such as antimalarial drugs. Favism does not occur in mild G6PD deficiency because the chemical oxidants present in the fava bean are not strong enough to damage red blood cells when G6PD activity is only minimally affected. Because of this, descendants of black Africans with mild G6PD deficiency can eat fava beans with impunity.

Why is G6PD deficiency more severe in males than females? This pattern is very typical of genes found on the human X-chromosome; so-called X-linked or sex-linked traits. The X-chromosome is the genetic Achilles heel of men. Women have two X-chromosomes, and therefore two copies of every gene on the X-chromosome, whereas men have only one. This makes men, as I will describe below, much more prone to inherited disorders caused by problems in genes on the X-chromosome.

Boys who inherit an X-linked trait can only get it from their mothers, since mothers provide their sons with an X-chromosome while their fathers provide them with a Y-chromosome. Girls inherit an X-chromosome from each parent. A father with an X-linked disorder will pass the trait down to all of his female offspring, who, in turn, will pass it down to half her sons. Sex-linked disorders are examples of inherited conditions that can skip a generation and be passed down from grandparent to grandson.

An important biological conundrum is how organisms compensate for the two-fold difference in the number of X-chromosomes (and the genes located on them) between males and females since a discrepancy in chromosome number usually leads to severe congenital problems. For example, having an extra copy of chromosome 21 causes Down's syndrome. Consequently, if women expressed the 2,000 genes present on each X-chromosome at twice the level as men, the sexes would be more strikingly different than we already are. Nature's solution to this "gene dosage" problem—i.e., females having twice the number of X-linked genes as males—is to compensate by inactivating one of the X-chromosomes in every cell in females. Chromosomal inactivation ensures that only one set of X-chromosome genes is active in any given cell for the duration of the cell's life. X-chromosome inactivation is passed down during mitosis (cell division); whichever chromosome is inactivated in a particular cell will remain inactive in its descendants. The process is completely random. One X-chromosome may be inactivated in one cell; the other in another cell. Thus, despite the two-fold disparity in the number of X-linked genes, the level of activity is essentially the same for both genders.[9]

The random inactivation of one X-chromosome in every cell in females is referred to as Lyonization, after Mary Lyon, a British scientist who first proposed this hypothesis in 1961 based on a series of observations made over the years by other scientists, and in the context of her own work on determining the cause of the curious color pattern found in the coats of female members of some mouse species, analogous to the coat patterns cat lovers admire in female calicos. Male calico cats have either orange or black coats; but only females can display a mosaic pattern with both colors. A gene for coat color is found on the X-chromosome, and there are two versions of it in the calico cat. One version codes for a black coat, and the other codes for an orange coat. White fur is also seen in these cats. This is caused by an albino gene on another chromosome and, if active, produces white fur regardless of which X-linked color genes are present. Because the black and orange color variants are found in a gene on the X-chromosome, male calico cats have either black or orange coats depending on the color of the mother's coat (remember that X-linked traits in males come from their mothers), and females with parents of the same color, will also inherit their single color pattern. There can be patches of white fur as well. This occurs in hair follicles that express the albino gene. However, if the coat of one parent of a female calico is of a color different from that of the other parent, her coat will contain a mosaic pattern of black and orange, mixed with patches of white. The mosaic pattern is due to

---

[9]A recent study by scientists at Pennsylvania State University and Duke has questioned this neat gene dosage solution. It turns out that many of the genes on the inactive X-chromosome are, in fact, somewhat active. So, for dozens, perhaps hundreds of genes, the level of expression in females exceeds the level of expression in males, perhaps contributing to some of the behavioral and biological differences that exist between the sexes.

the random inactivation of either the orange coat gene variant or the black coat gene variant in cells destined to become tufts of hair follicles, producing black fur growing from some follicles, and orange fur from others. On average, half of the coat will be one color and half the other color, creating a mosaic pattern. The white portion of a multicolored female calico is due to the lack of expression of either pigment gene, which occurs randomly in follicles that express the albino gene.

X-chromosome inactivation is due to a variety of factors including the effects of an X-linked gene called XIST (which stands for X-inactive specific transcript), which was discovered in the early 1990's by Dr. Hunt Willard. XIST expression results in the synthesis of an RNA molecule that shrouds an entire X-chromosome, preventing gene expression from taking place, rendering nearly the entire chromosome inert and useless. The precise mechanism by which XIST silences the X-chromosome has not yet been completely established. Another mechanism of X-chromosome inactivation is through a chemical modification of DNA called "methylation," which shuts down gene expression (more on this in chapter 14).

Random inactivation of one or the other X-chromosome is not just a biological curiosity that affects the coat color of cats. It also has consequences for humans with X-linked inherited disorders. If a mother carries a mutation on one of her X-chromosomes, she will transmit it to her daughters and sons with a 50 percent probability. However, the daughter can inherit an X-chromosome from the father that contains a normal copy of the gene. The son, who can only inherit a Y-chromosome from the father, will express the mutant gene inherited from the mother in every cell, since he has no other X-chromosomes. However, a cell in the daughter may or may not express the mutant version of the gene, depending on which X-chromosome was randomly inactivated in that particular cell. Thus, in one cell, she may inactivate the maternal X-chromosome and express the normal copy of the gene inherited from the father; another of her cells may inactivate the paternal X-chromosome and express the abnormal gene inherited on the maternal X-chromosome. Statistically, in an organ consisting of hundreds of billions of cells, half will inactivate the maternal X-chromosome and half the paternal X-chromosome.

The end result is 50 percent expression of the normal gene and 50 percent expression of the abnormal gene.

Expression at the 50 percent level is usually sufficient to protect girls from developing most severe X-chromosome-linked diseases, such as hemophilia. From the point of view of X-linked inherited disorders, men are unquestionably the weaker sex. Thus, Queen Victoria of England was a carrier of a hemophilia gene on one of her X-chromosomes but was unaffected by it during her reign as the British monarch for half a century, whereas her great grandson, Aleksey, the son of Czar Nicholas II and Alexandra of Russia, suffered because of it throughout his short life, debilitated by spontaneous hemorrhage into the joints, one of the more serious bleeding problems that occurs in hemophilia.

As in hemophilia and most other X-chromosome-linked disorders, girls with G6PD deficiency fare much better than boys. Because of random X-chromosome inactivation, girls who inherit one abnormal and one normal G6PD gene will have two populations of red blood cells. One, accounting for half the total number of red cells in the body, will express low levels of the G6PD enzyme, similar to that expressed in G6PD-deficient boys, and a second population with normal levels of G6PD. Thus, in fava-sensitive girls, half as many red blood cells will be destroyed as will be in boys, making them less prone to develop severe symptoms. It was random X-chromosome inactivation in girls, and not laziness in boys, that accounted for the gender differences in developing severe fatigue during the fava bean harvest.

The mutations responsible for G6PD deficiency have spread throughout the world by Darwinian selection because of a protective advantage against *falciparum* malaria. The highest rates of G6PD deficiency are found in the world's most malarious areas, or in regions where malaria used to be the most significant cause of death. Genetic studies indicate that some of the more common G6PD mutations are approximately five to ten thousand years old, corresponding to the time that slash-and-burn agriculture spread throughout the ancient world, which heralded the proliferation of malaria.

How G6PD confers resistance to malaria is still a mystery. There are studies showing that *Plasmodium falciparum* grows poorly in G6PD-deficient red blood cells. In women with one normal G6PD gene and one mutant version, *Plasmodium falciparum* growth is lower in the G6PD-deficient red blood cells. Other studies show that there is no difference in the rate of infection by *Plasmodium falciparum* in the two kinds of red blood cells, but do suggest that infection triggers an oxidative stress that results in the rapid removal of infected, G6PD-deficient red blood cells from the circulation, before *Plasmodium* merozoites can multiply and infect other cells, thus decreasing the chance that fatal cerebral malaria will occur.

There have been reports that the intracellular growth of *Plasmodium falciparum* is hindered when G6PD-deficient red blood cells are exposed to isouramil, one of the oxidizing chemicals present in fava beans. Other studies show that oxidants reduce *Plasmodium falciparum's* capacity to invade G6PD-deficient red blood cells. These findings suggest the intriguing hypothesis that fava beans may actually contribute to the protective effect against malaria afforded by having G6PD-deficient red blood cells. Thus, the cultivation of *Vicia faba* throughout the Mediterranean region and the Middle East may have come about because of its high concentrations of *Plasmodium*-thwarting oxidizing chemicals, as well as for its taste and nutritional value.

The spread of G6PD-deficiency genes throughout the malaria-endemic world is an example of a balanced polymorphism, a mutation that, though deleterious to some, increases in the population by natural selection through a strong survival advantage for others. Although the mortality from favism may have been about 10 percent in males with G6PD deficiency before the modern era, in evolutionary terms, it is more important that the condition enhances the reproductive potential of a G6PD-deficient female by reducing her mortality from malaria, even at the expense of sacrificing a few males. The mathematics of Darwinian natural selection can be harsh. If the balance sheet shows that 1 percent of males in a community will die of favism (the 10 percent rate of G6PD deficiency multiplied by the 10 percent mortality from favism), while the 10 percent of females who carry a single

G6PD deficiency gene gain a striking survival advantage against *falciparum* malaria, G6PD deficiency will win and the defective gene will expand in the population, despite the negative consequences to males.

The cold math of natural selection that led to the rise of G6PD deficiency and other genetic disorders, is an example, in the words of the film director Werner Herzog, of the "monumental indifference of nature."

CHAPTER 5

Neither Hurry, Curry nor Worry: Blood Type and Disease

Most people do not know their blood type, unless they are frequent blood donors or, as described later in this chapter, citizens of Japan. It is widely assumed that when blood is drawn for routine analyses, blood typing is part of the package; in fact, it is not. Blood typing is requested as part of a blood test only to find a matching donor for blood transfusion or, in the case of a pregnant woman, to determine whether a potential mismatch exists between maternal and fetal blood that can result in the potentially fatal destruction of fetal red blood cells. However, many people have a passing knowledge of the major blood types (A, B, AB, and O) and the sub-categories Rh-positive and Rh-negative. People with rare blood types, such as AB positive Rh-negative, who are sought by blood banks as valuable donors, wear their unique biochemical marker like a badge of honor. You would be hard-pressed to find someone bragging about having a rare variant of another blood protein such as hemoglobin or serum albumin.

The ABO blood-type system was discovered at the turn-of-the-twentieth century by Karl Landsteiner (1868-1943), based on his finding that people have antibodies in their blood that cause red blood cells from other people to clump; agglutinate is the scientific term. The antibodies bind to antigenic substances on the surface of red blood cells–proteins and glycoproteins (proteins that have carbohydrates attached to them)–that cover the entire membrane of the red blood cell like the fuzzy surface of a tennis ball. While the general structure of these surface molecules is similar in every person, there is some individual variability. This reflects small differences in the genes that code for membrane proteins, or differences in the genes coding for the enzymes involved in the synthesis of complex carbohydrates that attach to membrane glycoproteins. This variability does not affect the oxygen-carrying function of red blood cells–type-A red blood cells are as effective a transporter of oxygen as type-B cells–but it is nevertheless significant, because blood-type differences can cause potentially fatal reactions if incompatible blood is administered during transfusions. This can occur if the blood bank makes a laboratory error; it happens very rarely, but it happens. It can also occur in people who have been sensitized to a large number of red blood cell antigens because they have received many blood transfusions.

Another situation in which blood type incompatibility occurs is when there is a mismatch between maternal and fetal blood. This can sometimes stimulate the production of antibodies in mothers that can cross the placenta and destroy the fetus' red blood cells, resulting in a condition known as erythroblastosis fetalis, previously, the most common cause of stillbirth in developed countries. Before the molecular basis for this problem was understood and preventive measures initiated, serious or fatal reactions caused by mismatched maternal and fetal red blood cells probably occurred tens of millions of times in human history.

Such blood-group incompatibility arose as an unintended consequence of nature's design, when diversity in the structure of the surface proteins on red blood cells evolved, in part, as protective mutations against malaria and other serious infections.

The ABO blood-type system is caused by a common variation in a gene found on chromosome nine that codes for an enzyme involved in the addition of sugar molecules to a protein attached to the cell membranes. The genetics of the ABO system is fairly straightforward. Everyone has an O gene (except for a few rare individuals), which results in an O-molecule core consisting of a chain of four molecules of complex sugars (fucose, galactose, N-acetylglucosamine, and glucose), which is attached to a protein base situated in the membrane of red blood cells and other cells. Type-A blood is caused by an A gene, which adds another sugar molecule (called N-Ac-D-galactosamine) to the O molecule. Type-B blood is caused by the B variant of the same gene, resulting in the addition of a different sugar (D-galactose). If you have both the A and B variant genes inherited from each parent, half of your O molecules will contain N-Ac-D-galactosamine and half will contain D-galactose, ending up with type-AB blood. If neither the A nor B variants are expressed, there is only an unmodified O molecule, which results in type-O blood.

People produce antibodies to different blood types because the glycoprotein coats found on type-A and type-B red blood cells are similar to those found on the surface of many types of bacteria that are ubiquitous in nature. These antibodies (also called agglutinins) will attack red blood cells that are recognized as foreign. Fortunately, the immune system usually inactivates B-lymphocytes that produce antibodies to proteins and glycoproteins present in that person's body. Thus, a person with type-A blood produces antibodies to type-B blood while those with type-B blood produce antibodies to type-A. People with the AB blood type produce neither antibody and therefore have no reaction against any other ABO blood type, which is why they are commonly called "universal" recipients; whereas type-O blood cannot be attacked by anti-A and anti-B antibodies thus making people with this blood type "universal" donors. This is why in an emergency (or in emergency room dramas on TV), when there is no time to identify a person's blood type because he or she is in shock, doctors scream out for "O-negative blood" (type O and Rh-negative).

Rh factor is a protein complex present on the surface of red blood cells of most people that is equally important in transfusion medicine. Rh blood type is controlled by a different set of genes from those responsible for the ABO complex, and its genetics is rather complicated; there are more than two dozen different forms of the Rh molecule. For simplicity, assume that there are only two Rh forms: positive (also known as "D"), found in approximately 85 percent of the population, and negative (also known as "d"), found in 15 percent. In contrast to the ABO system, the human body produces no natural antibodies against Rh-positive red blood cells.

Antibodies to Rh are produced only when an Rh-negative person has been exposed to Rh-positive blood cells.

The practical knowledge gained from studying the ABO and Rh systems has saved millions of lives over the past century through the ability to provide compatible blood and blood products for transfusions, and has also prevented millions of stillbirths.

Landsteiner's landmark discoveries occurred at a time when the true function of blood as the vehicle for delivering oxygen and nutrients to every cell in the body had been known for only a relatively short period. Before the 19th century, blood was viewed as a medium that carried personality traits and the soul. In fact, it was believed that personality traits could be transferred by drinking blood, a habit practiced by people in many different cultures. The Romans used to drink the blood of slain gladiators to acquire their strength and courage. The inventor of dynamite, Alfred Nobel, believed in the rejuvenating qualities of giraffe blood. The Aztecs, who learned how to harness the pharmacological and culinary properties of cocoa beans, viewed chocolate as a symbol for human blood. They created a frothy chocolate drink mixed with spices and red dye to evoke the blood of sacrifice victims and gave the concoction to their victims before torturing and killing them.

Blood is, in fact, an excellent nutritional source, as is evident from the diversity of creatures who consume it as their sole or principal food, including vampire bats, mosquitoes, fleas, and ticks, sometimes with devastating effects on humans. Blood's constituent proteins, carbohydrates, and lipids are broken down by the digestive system of blood drinkers into the same absorbable nutrients of any other food, and it is the richest natural source of iron. Hemoglobin is the most abundant protein and the primary nutrient in blood. Blood is not only a food source for mosquitoes and ticks but is also consumed by the parasites they sometimes carry, such as *Plasmodium* and a protozoan that causes a rare parasitic infection of red blood cells called babesiosis, which is carried by the same tick that causes Lyme disease. These intracellular parasites live off hemoglobin, consuming blood one cell at a time from the inside out.

Blood is still a dietary staple for some people. The Masai, a pastoral tribe in Kenya and Tanzania, have a long tradition of drinking a highly nutritious mixture of milk and cow's blood. Some European cultures have a taste for sausages made with animal blood.

A vestige of the belief that blood transmits personality traits survives in modern times; consider the expression "It's in his blood"–referring to a person's natural proclivity for something, whether it is musical talent or homicidal mania. Although few Westerners know their own blood type, more than 90 percent of the people do in Japan, where certain personality traits are attributed to specific blood types: people with type-O blood are considered to be calm and self-confident; those with type-A blood are cautious, well-mannered, and self-sacrificing; type-B blood makes people cheerful, flamboyant, and outgoing; and type-AB blood endows people with type-B traits on the outside and type-A traits on the inside. In South Korea, where interest in blood type and personality has been passed down from the Japanese, quiet, reserved type A women are warned to avoid dating type B men, who are viewed as cads. Blood type has also been used to explain the four classes of feudal Japan–warriors being type-O, farmers type-A, tradesmen type-B, and artisans type-AB. Although most Japanese regard a possible association between blood type and personality with the same frivolousness that a Westerner might have toward a horoscope, asking "what's your blood type" instead of "what's your sign" at a party or on a first date, some take it quite seriously. In the 1920s, during the early stages of the Japanese military expansion, blood typing of military personnel was used to help match recruits to their most suitable military roles. In modern Japan, some companies still utilize it to determine an employee's aptitude for certain jobs, and some schools will base their choice of different teaching strategies for kindergartners on the children's blood types. Instead of referring to daily astrological horoscopes, some Japanese search for good fortune in their blood-type horoscope published in newspapers and women's magazines.

The belief that blood carries personality traits provided the rationale behind the earliest attempts at intravenous transfusion. The idea of treating a patient for an illness by injecting a substance into a vein or artery was inspired by the landmark discovery in 1628 by William Harvey (1578-1657) that blood circulates throughout the body. Further research on circulation of the blood was pursued by Thomas Willis (1621-1675) and his circle of students in Oxford and London in the 1650s and '60s, including Sir Christopher Wren (1632-1723), Richard Lower (1632-1691), Robert Hooke (1631-1703), and John Locke (1632-1704). Wren, in addition to being the architect of St. Paul's Cathedral (and more than 50 churches in London) as well as homes, hospitals, and libraries, was also a mathematician, astronomer, physiologist, and medical experimenter. In fact, he was one of the first to attempt intravenous injections by administering wine and opium into the veins of dogs using quills. Wren's experiments led the way to the earliest recorded successful blood transfusion experiments by Lower, who in February 1665 performed the first recorded dog-to-dog transfusions.

The first known successful blood transfusion in humans occurred in January 1667, performed by Jean Baptiste Denis, a Frenchman who transfused blood from docile animals—lambs and calves—to mentally ill people believing that it would help calm them. One of his patients was the son of the prime minister of Sweden, who died from an incompatibility reaction soon after the procedure. This public setback did not prevent the transfusion of other patients. One was Antoine Mauroy, a 34-year-old man who was psychotic and violent, whom Denis tried to calm with transfusions of calf blood. Because Mauroy seemed to improve after the first transfusion, he was given a second one, this time using a larger quantity of blood, after which he became feverish, developed severe back pain, and began to vomit; and the next morning, despite feeling a little better, he passed a very dark urine that was described as the color of urine mixed with soot—all signs of a transfusion reaction, from which Mauroy eventually recovered. However, within a few months, his mental state had gotten worse and his desperate wife urged another transfusion, and Denis obliged. Unfortunately, this second procedure caused a fatal transfusion reaction. Following an all too modern practice, Mauroy's wife sued Denis, accusing him of poisoning her husband, and the good doctor was charged with murder. Although Denis was ultimately acquitted, he was prevented from conducting further transfusion experiments, and ultimately, the procedure was banned by the French Parliament.

Experimental transfusion treatment also took place in England. Richard Lower, having pioneered the procedure, and hearing about Denis's work, transfused a mentally ill man named Arthur Coga with lamb's blood in December 1667. His overly enthusiastic report to the Royal Society suggested to some in the audience that the investigator himself may have been "cracked in the head," and the report was dismissed by the medical establishment. Eventually, the British Parliament and other European governments followed the French lead and banned transfusions, effectively halting transfusion experiments for nearly two centuries.

So ingrained was the belief that blood was the source of behavioral traits that, amazingly enough, the possibility of using transfusions for the purpose of replacing blood lost in an acute injury or trauma never occurred to early investigators. It was not until 1818 that blood transfusions were considered for treating acute hemorrhage by a London obstetrician, James Blundell, who was motivated by having seen women bleed to death from postpartum hemorrhage, which was then the second most common cause of childbirth mortality (the first being infection). After Blundell performed experiments using same-animal transfusions, he successfully transfused a hemorrhaging woman with some blood from her husband's arm in 1818 and, between 1825 and 1830, attempted 10 more human-to-human transfusions. Five of these were apparent successes, roughly the rate one would predict using the law of probability if transfusion is performed without knowing the donor or recipient blood type. The great 19th-century American physician William Steward Halsted used his own blood to effectively transfuse his sister, who at the time was near death from a severe postpartum hemorrhage.

During the late 19th century, because of the unpredictable nature of transfusion reactions, the infusion of human blood was considered so dangerous (a correct assumption at that time) that it could only be recommended as a last resort in the treatment of acute blood loss. Consequently, the procedure never caught on as a general medical practice.

The cause of transfusion reactions was eventually determined by experiments that began in the late 19th century, when Leonard Landois and Emil Ponfick in 1874 showed that human serum contains a substance, an agglutinin, that causes the red blood cells of animals to clump. However, because the experiments were performed on sick people, the importance of the observation was overlooked; the presence of the agglutinins was attributed to disease.

Landsteiner carried out a simple experiment to show that agglutination was a physiological process, and not a consequence of disease. He arrived at this conclusion by separating the red blood cells and serum of healthy co-workers, and by testing different combinations of the two. Landsteiner found that normal serum contained agglutinins that caused red cells from some unrelated individuals to clump. Based on experiments conducted on a relatively small number of subjects, he was able to quickly identify the three most common groups, which he called A, B, and O. It took a few more years before the rarest group, AB, was discovered by Decastrello and Sturli. Using agglutinins to determine blood type, Landsteiner and other researchers were able to prevent most transfusion reactions. These simple experiments provided doctors with ability to safely administer blood to save the lives of thousands of soldiers wounded during World War I.

By the 1920s, tens of thousands of blood transfusions had been performed safely in United States, revolutionizing the fields of trauma medicine and surgery. Landsteiner hailed the remarkable safety record of transfusions resulting from his discovery of blood types in the speech he gave to accept his Nobel Prize in 1930. In 1,467 successive transfusions at Bellevue Hospital in New York City, he pointed out, there were only two deaths, one of which was caused by a laboratory error.

Landsteiner stands as one of the giant figures of early 20th-century medicine. In addition to his work on blood groups, he also established the viral etiology of polio, a discovery that ultimately led to the vaccines for polio developed by Jonas Salk and Albert

Sabin during the 1950s. It is a peculiarity of history that even though Landsteiner's discoveries have benefitted hundreds of millions of people, almost no one knows his name today, whereas most people in modern society have heard of his contemporary and countryman Sigmund Freud, many of whose theories about psychiatric disorders were widely accepted for half a century without scientific scrutiny but have since been largely superseded by modern neurobiological findings and other behavioral theories. Admittedly, though, red blood cell agglutination does not carry the same luster as unresolved childhood sexual conflicts and dream interpretation. Landsteiner won his Nobel Prize in 1930, the same year as Sinclair Lewis's Nobel Prize in literature. During a conversation in Lewis's hotel room, the writer apparently took off his clothes, got into his pajamas and fell asleep, literally in mid-conversation with Landsteiner, who was describing his work. No doubt, Lewis's chronic state of intoxication, as well as Landsteiner's topic of conversation, played a role in the author's drift into unconsciousness. Despite a wonderful testimonial by Lewis during the ceremonies, Landsteiner held a grudge against the writer, a feeling shared by many of the friends and acquaintances the abrasive Lewis managed to antagonize during his drunken years.

The Nobel Prize was a bit of a distraction for Landsteiner, who was a very private man. He cared for his mother for her entire life, even postponing marriage until after she died, when he was in his late forties. Landsteiner kept her death mask above his desk for his remaining years. He was dedicated to his work, and his sense that the growing anti-Semitism in Austria would hinder his career may have prompted his conversion from Judaism to Christianity during his early medical training. Distancing himself from the religion of his ancestors took an extreme, somewhat paranoid twist in the late 1930s, after he had moved to the Rockefeller Institute in New York City: he sued the publishers of a *Who's Who* of prominent Jews to have his name removed from the publication, fearing that it might prompt anti-Semitic hostility against not only himself but his wife and son too. Although he dropped the suit, his reputation was tarnished a bit by the incident.

Landsteiner also played an instrumental role in the discovery of other blood groups, most notably, the Rh system. It was initially described in 1939 in a woman who was recovering from the birth of a stillborn child with erythroblastosis fetalis. She lost a lot of blood while delivering the dead baby and was given transfusions provided by her ABO-compatible husband. Yet, she developed a severe transfusion reaction. The woman's serum was found to contain an agglutinin that reacted with her husband's red blood cells, as well as with red cells from 85 percent of other humans, but not with her own. Landsteiner and his colleagues proposed that the mother became sensitized to an unknown substance on her husband's red blood cells, which was transferred across the placenta and destroyed her fetus's red blood cells. The same agglutinin was found in serum obtained from rabbits and guinea pigs immunized with red blood cells from rhesus monkeys, and so it was called "Rh" factor.

There are no natural substances that generate Rh agglutinins in animals. People who are Rh-negative produce these antibodies only when exposed to Rh-positive blood. This occurs in two circumstances: the accidental transfusion of Rh-positive blood into an Rh-negative recipient, and pregnancy. The combination of an Rh-negative mother and an Rh-positive father—a match that occurs in about 10 percent of all couples—turned out to be the common denominator in nearly all cases of erythroblastosis fetalis in Western countries. Depending on whether the father has one or two copies of the Rh gene (actually the "D" gene component of the Rh gene complex), either half or all the fetuses growing in the uterus of Rh-negative mothers will be Rh-positive. During delivery, a small amount of fetal blood gets into the mother's circulation. Exposure of an Rh-negative mother to her baby's Rh-positive blood will sensitize her immune system to produce Rh agglutinins. The first Rh-positive fetus growing in an Rh-negative mother does not usually experience problems; the most important period of maternal sensitization occurs during delivery, when fetal and maternal blood mix. However, if she conceives a second Rh-positive fetus later on, there is a sufficient quantity of Rh-agglutinating antibody present in the mother to cross the placenta and destroy the developing fetus's red blood cells, causing erythroblastosis fetalis; firstborn Rh-positive children are immunological Cains, sensitizing their Rh-negative mothers to destroy their future siblings.

Erythroblastosis fetalis caused by Rh incompatibility can now be entirely prevented using a biological anti-missile system. Pregnant women who are Rh-negative receive injections of an antibody called RhoGAM during pregnancy and at the time of delivery. RhoGAM is a highly concentrated Rh agglutinin that attaches to and sequesters fetal Rh-positive red cells that cross into the mother's blood during delivery. Fetal Rh-positive cells coated with RhoGAM are rapidly destroyed by the innate immune system before the cells have time to sensitize B-lymphocytes, which would otherwise have produced large quantities of Rh-agglutinating antibody. The theoretical framework and clinical application behind this elegant immunological trick was developed by doctors Vincent Freda, John Gorman, William Pollack, Ronald Finn, and Cyril Clark in the early 1960s. Although the idea was initially condemned, a well-designed clinical trial in 1964 firmly established RhoGAM's effectiveness in eradicating erythro-blastosis fetalis caused by Rh incompatibility.

Sometimes a mother with type-O or type-B blood destroys the red blood cells of her fetus when they are type-A, regardless of Rh type. Since each of the three major blood types (O, A, and B) are so common, it seems as if this should occur quite often, considering that antibodies to type A are found in everyone with type-O and type-B blood. Humans are lucky. The naturally occurring antibodies produced against type-A and type-B red blood cells are of the IgM variety, which do not cross the placenta; only IgG antibodies cross over from mother to fetus. Rh agglutinin is an IgG antibody. If antibodies to type-A red blood cells were consistently of the IgG variety, there would have been hundreds of millions of fetuses and newborns lost throughout human history.

Although the ABO and Rh systems are the most powerful determinants of blood group compatibility for transfusions, there are dozens of different proteins and glycoproteins present on the surface of the human red blood cell that can trigger an immune response and transfusion reaction. Because such reactions tend to be less severe than ABO- and Rh-incompatible reactions, these other surface molecules are sometimes referred to as "minor blood groups." One of the more important minor blood groups is the called "MN," which is caused by genetic diversity occurring in the gene that codes for a glycoprotein called glycophorin A. Another is the "Duffy" blood group system, named for a hemophiliac who developed a new red blood cell agglutinin after receiving multiple transfusions.

The term "minor blood group" is a misnomer. These molecules provide important functions to the red blood cell and are "minor" only in the sense of being less important in transfusion medicine than ABO and Rh. For example, glycophorin A and related proteins, glycophorins B and C, help generate a zone of negative charge that surrounds red blood cells. This zone of negativity helps keep red blood cells from sticking to each other, like magnets that repel when the same poles are face to face, and helps prevent red blood cells from attaching to the inner walls of blood vessels. Other minor blood group determinants serve physiological roles as ion channels and transporters, and their loss from the cell has adverse effects on the function of the red blood cells.

Paradoxically, the function of the major blood group determinants in red blood cell physiology remains elusive. The ABO proteins are not critical for life. There are a few rare individuals who are born without the ability to make A, B, or O molecules, a condition that does not pose any significant health problem, except being the only people in the world who cannot be transfused with type-O blood. They have no alterations in the shape or size of their red blood cells, and those cells' ability to deliver oxygen to tissue is not impaired. These observations would seem to suggest that the ABO system is something of a vestige in the human body, like the appendix or earlobe, and can be eliminated without any untoward effects. Similarly, there is no obvious difference in red blood cell function between Rh-positive and Rh-negative red blood cells. What is the purpose of such a diversity of proteins on the surface of red blood cells? Why has nature provided us with this apparently counter-evolutionary system, destroying millions of developing fetuses throughout human history (due to maternal-fetal Rh incompatibility), while having no detectable effect on the main function of red blood cells–the transport and delivery of oxygen? The answer, at least for some of these blood-type differences, lies in the biology of

Plasmodium attachment and entry into red blood cells, critical events in the infectious life cycle of the parasites that cause malaria.

The affinity that microorganisms have for particular cells, so essential to the infectious process, is based on their ability to exploit the normal physiology and microscopic anatomy of the surface membrane of host cells. The architecture of cell membranes is as intricate and detailed as a Gothic church. Millions of molecules consisting of hundreds of different proteins cover the cell surface, creating a three-dimensional landscape of molecular scaffolds, edifices, and bridges, quite unlike the smooth, circular images of cells depicted in elementary biology texts. Surface-membrane proteins are the primary means by which cells communicate with one another and relay information. They are the way stations that connect the outside world of the extracellular space–with its hormones, growth factors, nutrients, and salts–to the cell's inner sanctum. Nearly every substance that makes its way into the cell does so via a membrane protein specifically designed to accommodate that substance: potassium channels for potassium, calcium channels for calcium, and so on. Substances that do not enter cells, such as certain hormones and growth factors, relay information by binding to a specific cell-surface receptor, which then transmits signals to the inner workings of the cell–a process so ubiquitous and complex it permeates all aspects of cell biology, from the ability to perceive odor, vision, and sound to the development of cancer. The process of information transfer in cells is referred to as "signal transduction." (more on signal transduction in later chapters).

Infectious microorganisms that grow inside of cells–including all viruses, many bacteria, and *Plasmodium*–have acquired, as a means of gaining entry into the cell, the ability to exploit the surface molecules used by host cells in their day-to-day physiology. They do this by first attaching to the surface molecules using specific receptors encoded in their genomes. Microorganisms that enter cells as part of their life cycle do so for a variety of reasons, none of which are mutually exclusive. One common reason is that the cell can provide microorganisms with nutrients, such as hemoglobin. Some invading microorganisms, such as viruses, may not have the necessary machinery to multiply on their own and have to parasitize the biochemistry of host cells in order to replicate and synthesize protein. A third reason is that the intracellular environment can act as a sanctuary from attacking antibodies.

After attaching to the cell membrane, microorganisms enter the cell using one of several different strategies. The outer coat of some viruses fuses with the host membrane, enabling the viral genome to empty into the cell. Sometimes, a microorganism will fool a host cell into creating a membrane bud that surrounds it, after which it is engulfed by the cell, bringing the entire microorganism into the intracellular confines, a process known as endocytosis. By stimulating endocytosis and membrane fusion, invading microbes exploit fundamental processes of life, the same processes that occur when a sperm enters an egg and fertilizes it, and when a phagocyte engulfs a microbe. Indeed, microbial invasion by endocytosis has been instrumental in the development of life on Earth. About 1.5 billion years ago, a primitive eukarotyic cell was invaded by a single-celled creature belonging to the Archea family, and a symbiotic relationship developed: the host cell provided some proteins that stabilized the structure of the invading microbe, while the invader provided enzymes that increased the production of the energy molecule ATP. These invading microbes eventually developed into mitochondria, the energy-producing organelle found in every eukaryotic cell on Earth, except red blood cells. Another microbial invasion that radically altered life on Earth resulted in the development of photosynthesizing plants. This occurred when an ancient eukaryotic cell was invaded by a microbe of the Cyanobacteria family. These microbes ultimately evolved into chloroplasts, the plant cell organelles that contain chlorophyll, which is essential for photosynthesis, the enzyme-driven chain reaction that converts sunlight and carbon dioxide into sugar and oxygen.

Another strategy invading microorganisms use to enter cells is to attach to a specific receptor and secrete an enzyme that digests a small hole in the host cell membrane through which it can enter. *Plasmodium* uses this mechanism to attack red blood cells.

Each *Plasmodium* species has developed its own unique attachment strategy. In human malaria, *Plasmodium vivax* attaches to the Duffy blood group protein to gain entry while *Plasmodium falciparum* attaches to glycophorins.

The fact that *Plasmodium vivax* requires Duffy protein in its invasion plan has led to a remarkable Darwinian survival mechanism: nearly 100 percent of West Africans and their New World descendants, including a majority of African-Americans, have the Duffy-negative blood type, which makes them completely resistant to *vivax* malaria, whereas Duffy negativity is extremely rare in other ethnic groups. The Duffy-negative blood type has been so successful that *vivax* malaria has been eliminated in much of West Africa, a small oasis of relief from one of many infectious diseases (unfortunately, the variant does not protect against other types of malaria). In one study of more than 1,000 Duffy-negative subjects from Gambia, not a single blood sample was infected with *Plasmodium vivax*, yet the other malaria-causing protozoa were well represented. Decades ago, in an era when human experimentation did not follow the same ethical restrictions that govern modern research, some investigators in the United States intentionally infected volunteers with *Plasmodium vivax*, and it was noted that most African-Americans did not develop malaria. Imagine an Army general with knowledge of this tendency ordering only African-American soldiers into a battlefield where the fighting forces have to contend with *vivax* malaria as well as bullets, a crafty but racist strategy.

The genetic variant responsible for Duffy-negative red blood cells is one of the most clever mutations to emerge from natural selection against an infectious disease, something the best genetic engineers would have been proud to design. Duffy protein is expressed in many cell types and functions as a binding site for hormones and growth factors involved in inflammation, an important function that might handicap Duffy-negative people. However, the mutation that has emerged as the most important variant resulting in Duffy-negative red blood cells is designed to guarantee preservation of Duffy function in other cells. The mutation is in the Duffy gene "promoter," the master regulator of gene expression, and it affects the binding ability of a red blood cell–specific member of the gene-activating family of proteins called GATA. The mutation prevents GATA-induced Duffy gene activation and Duffy protein synthesis only in red blood cells; expression is preserved everywhere else in the body, since the promoter maintains its ability to respond to gene-activating proteins expressed in other cell types. The mutation is a perfect manipulation of the molecular biology of a gene and the biology of a parasite. Not only is there complete protection against *Plasmodium vivax* attachment, but the active function of Duffy protein in other cells of the body is unaffected.

From the perspective of population genetics, there is something a bit unusual about the resistance to *Plasmodium vivax*. The selection for Duffy negative blood type involves a disease, *vivax* malaria, which is usually not fatal. When selective pressure is exerted favoring a particular genetic variant, the catalyst is an improvement in reproductive fitness, helping a species survive to reproductive age so that its genes are passed down to the next generation. Each of the malaria-protecting mutations discussed in this book has emerged because of resistance to *Plasmodium falciparum*, the only malaria parasite consistently fatal in children. It is a mystery why one of the most striking examples of Darwinian selection against a disease has developed for a form of malaria that is rarely fatal. Perhaps it was a combination of factors that caused early encounters with *Plasmodium vivax* to be more destructive than it is today; lack of natural immunity, malnutrition and the debilitating effects of other infectious diseases are possible explanations. Whatever the reason, long ago, West Africans with red blood cells containing a Duffy-glycoprotein coat, *Plasmodium vivax's* point of entry for invasion, were wiped out.

Mutations in the genes that code for glycophorins have been found to protect against invasion by *Plasmodium falciparum*. Genetically engineered mice that are born without glycophorin A are resistant to infection with mouse malaria. The best example in humans is found in the malaria-endemic coastal areas of Papua New Guinea. There, 50 percent of the population has inherited a defective glycophorin C gene which helps prevent *Plasmodium falciparum* from invading red blood cells.

The role of the ABO blood groups in malaria infection is a matter of controversy. Some epidemiological studies in Africa show no relationship between ABO groups and malaria. However, several studies suggest that people with type-A blood have a higher risk of developing cerebral malaria. No major connection between malaria and Rh group has been made.

Although a large portion of this chapter has been devoted to discussing the major blood types, ABO and Rh, their relationship to malaria is tenuous. In conferring malaria protection to humans, the "minor" blood types are much more important.

Although the malaria connection is weak, the ABO blood group system plays a role in other diseases; peptic ulcers for one. People with type-O blood are more prone to getting them. During my medical training in the mid 1970s, doctors believed that peptic ulcers were caused by stress and excess secretion of acids. The only medicines available at the time were antacids that buffered stomach acid and drugs that blocked acid secretion. These medications relieved ulcer pain and promoted healing. However, ulcers frequently recurred and some had a tendency to bleed, sometimes as vigorously as a gunshot wound. Once, a homeless man was brought into the emergency room where I was clerking as a medical student. He was so filthy that the nurses brought him into a room to get washed, even before taking his pulse and blood pressure. In the middle of his cleaning, he lost consciousness. We eventually figured out that he had an actively bleeding ulcer and had lost more than half his blood volume. But that knowledge came too late–the patient had died.

In retrospect, ulcers kept recurring because we were only treating the symptoms and not the underlying cause.

Twenty years ago, two investigators, Barry Marshall and J. Robin Warren, reported in the medical journal *Lancet* the discovery of a novel bacterium, *Helicobacter pylori*, from biopsy samples taken from the stomach and duodenum of patients with gastritis (an inflammation of the stomach) and peptic ulcers. They proposed that the painful and sometimes fatal problem of peptic ulcer disease was really an infectious disease. Like many revolutionary departures from conventional medical wisdom, their hypothesis was initially met with skepticism. In a throwback to courageous researchers of the past who deliberately exposed themselves to potentially deadly microbes in order to establish their role in infection, Marshall went so far as to ingest *Helicobacter pylori*. He came down with a case of painful gastritis, a harbinger of peptic ulcer disease, and submitted to an upper GI endoscopy to prove it (a flexible scope is passed from the mouth into the stomach and duodenum, which are surveyed and biopsied). He cured himself with a course of antibiotics, becoming the first of millions to intentionally rid themselves of a *Helicobacter pylori* infection. Several more years of study proved beyond a doubt that the primary cause of peptic ulcer disease was *Helicobacter pylori* rather than personality type, stress, or spicy food–hurry, worry, and curry, as the adage goes. Marshall and Warren were awarded for their tenacity and self-sacrifice with the 2005 Nobel Prize in Physiology or Medicine.

Helicobacter pylori is transmitted from person to person by vomit and fecal contamination of food and water, which is why infection is more common in underdeveloped countries. This bacterium has been around long enough for different variants to emerge in human populations, a characteristic that geneticists and anthropologists have used to study human migration. They have found, for example, that DNA isolated from the remnants of a strain of *Helicobacter pylori* in ancient Chilean mummies differs from the European type that infects modern-day Central and South Americans. Like the conquistadors who overthrew indigenous tribes and drove all but a few to extinction by genocidal warfare and infection with the European diseases smallpox and measles, it appears that the foreign *Helicobacter pylori* strain conquered its local counterpart as well.

The association between ulcers and type-O blood group is difficult to explain, both physiologically and from an evolutionary perspective. In addition to being produced in developing red blood cells, ABO glycoproteins are also found in the gastrointestinal and respiratory tracts. However, it's not clear how the presence of these substances in the gut influences whether *Helicobacter pylori* lives as a harmless commensal bacterium, as it does in the overwhelming majority of infected people, or whether it punches holes in the stomach and duodenum. Furthermore, since *Helicobacter pylori* infection does not appear to influence reproductive fitness, it should not have an affect on the frequency of ABO blood types in the population.

A relationship between ABO glycoproteins expressed in the gut and infectious causes of diarrhea has also been proposed. Severe diarrhea can cause selective pressure in human evolution because it can produce circulatory collapse and death in small children (see chapter 7). One of the most common causes of infantile diarrhea is a virus called rotavirus. In one study of a rotavirus subtype (albeit a strain that infects cattle), the virus was found to bind to type-O red blood cells. It has also been reported that cholera infections are worse in individuals with type-O blood, which would have selected for blood types A and B; however, the evidence is weak. There are some studies that also show a relationship between ABO blood group and other causes of severe infectious diarrhea, such as enteropathogenic *E. coli* and the Norwalk Virus (NV), two of the major nuisances for foreign travelers and cruise passengers. People with type-B blood have been reported to have less severe infection with both *E. coli* and NV. The mechanism of the protective effect of type B is not clear, but an impact on the binding of these microbes to cells lining the gut, which affects their ability to invade the wall of the small intestine, is likely.

Some investigators have speculated that there may be a relationship between ABO blood groups and influenza. Influenza usually causes a severe but short-lived upper respiratory illness characterized by cough, fever, headache, and malaise. However, tens of thousands of deaths are attributed to influenza every year, mainly among the elderly, people with chronic illnesses, and infants. Occasionally, an extremely virulent strain of influenza emerges, capable of killing millions of people. The most serious was the great 1918 pandemic which spread across the world, aided by troops mobilizing for the final battles of World War I. Some studies have shown susceptibility to influenza to be related to blood groups B and AB, and others to O and A. These confusing findings could be due to the differences in the interaction that various influenza strains have with ABO proteins present in the respiratory tract; some strains may be thwarted by one blood group, others by a different one. However, most studies showing a relationship between influenza and ABO date from the 1970s and '80s–ancient history in the world of molecular genetics and virology–and the majority of studies were conducted by obscure researchers in the former Soviet Union.

Considering the possible association between different ABO blood types and a variety of infectious microorganisms, from *Helicobacter pylori* to influenza and Norwalk viruses, it is conceivable that selection for one blood type or another has occurred against many dangerous infectious diseases throughout human history, resulting in a fairly balanced frequency for each of the different blood types in the human community. After nearly a century of research, the ABO and Rh blood type/infectious disease connection is still unresolved. However, one clear association, albeit, an indirect one, between blood type and infectious diseases has been the rise of blood-borne pathogens, notably HIV and the viruses responsible for hepatitis B and hepatitis C. While Landsteiner's discoveries provided doctors with the ability to safely transfuse billions of units of blood over the years, it also resulted in the transmission of these blood-borne viruses to hundreds of thousands of people. Even after testing donor blood became routine in the mid 1980's, tens of thousands more were infected in poor countries, where screening was either unavailable, or unaffordable. In some cases, questionable use of transfused blood, corrupt political systems, and antiquated and unscrupulous medical practices conspired to help spread HIV and hepatitis. In Romania, for example, doctors transfused orphans unnecessarily under the scientifically invalid belief that it would bolster their health (when in fact, what they really needed was adequate nutrition and loving caretakers). Hundreds of orphans developed AIDS. Doctors failed to recognize the illness and how it was transmitted, despite the fact that the clinical aspects of AIDS and its means of transmission had been known for several years throughout the rest of the world, because they were kept ignorant about the emerging pandemic by the Stalinist-like dictator, Nicolae Ceausescu. The self proclaimed "genius of the

Carpathians," had effectively isolated the country from outside influence, including vital new medical information. By 1990, when Romanian doctors finally learned about AIDS, seven years too late, nearly 50 percent of all pediatric AIDS cases in Europe could be found in the Black Sea port city of Constanta, which has a population of only a few hundred thousand.

In China, unscrupulous blood traffickers have spread HIV through unsafe collection practices and by not testing for the virus to save on cost. Poor farmers who are paid to donate blood have been the major victims. Blood traffickers contaminated their donors by reusing dirty needles. They contaminated donors and recipients alike through the particularly depraved practice of mixing blood from multiple donors in tubs, siphoning off the plasma, which was sent to hospitals, then transfusing the remaining red blood cells back into donors in order to enable them to donate again more quickly.

Scientists will continue to search for more natural associations between blood type and disease as new infectious diseases emerge. Perhaps some discovery will be made on the effect of blood type on an emerging infection, or a virulent strain of influenza that caused devastation in the past, thereby providing new insight into why the differences of the major blood groups evolved. That microbe may now be lurking in a handful of people in Africa, Central America, or Asia, hidden among the more conspicuous problems facing the poor nations of the world, waiting its turn to battle humanity.

CHAPTER 6

Inhospitable Red Cells

When medical students learn how to draw blood, they are instructed to do so with the patient lying down or sitting comfortably in a chair to prevent the patient from falling with a needle stuck in a vein if he or she passes out. Many people get woozy at the sight of blood. But hematologists–doctors who specialize in diseases of the blood and bone marrow–regard blood as a thing of beauty. It is most magnificent when red cells are forced to burst open in a test tube when blood is suspended in a dilute solution of salt or sugar or just plain distilled water. This causes the hemoglobin to explode out of the cells, coloring the solution a deep shade of burgundy. Whether the sight of blood makes you feel sick or makes you think of a fine port, the colorful protein inside, hemoglobin, should be appreciated from an evolutionary perspective–its involvement in a survival-of-the-fittest struggle between animals and parasite-harboring mosquitoes that goes back millions of years. The need for hemoglobin as a primary food source for egg-laying female *Anopheles* seems a bit improbable. The insect had to evolve complex behavioral and physiological strategies, such as biting a host while he or she is asleep, and delivering a local anesthetic through its proboscis, to avoid being swatted dead by animals that outweigh it by a million-fold. It is as if, in a prehistoric world filled with a variety of food sources, the only mechanism that pregnant women developed for getting nutrients into their developing fetuses was to have them sneak up to and bite sleeping mastodons for a blood meal; strange pregnancy cravings indeed! The mosquito and the dangerous parasite it harbors have teamed up and learned how to harvest hemoglobin from a host of different vertebrates, their relationship molded by the co-evolution of hundreds of different genes, from those that direct feeding and nesting behavior in *Anopheles* to the ones that control digestive enzymes in *Plasmodia*.

But, mosquitoes and single-celled protozoa are not the only ones with genetic tricks in their repertoire. We humans have thrown own genomes into the mix to combat the deadly consequences of the marriage between *Anopheles* and *Plasmodium falciparum,* by generating hundreds of mutant versions of the genes responsible for making hemoglobin itself.

Hemoglobin is the complex protein that carries oxygen absorbed from the lungs for distribution to every cell in the body. It is one of the most abundant proteins in the body, and one of the most thoroughly studied proteins in nature having been probed and dissected from every scientific angle imaginable for more than half a century. It was one of the first proteins to be characterized at the three-dimensional level using X-ray crystallography, an analysis that revealed a twisted, tubular structure, like a long sausage link folding back on itself. The genes coding for globin– the protein portion of hemoglobin–were among the first to be cloned and to have their genetic structure determined. The globin genes are also targets for some of the most common inherited disorders of mankind; collectively, mutations affecting the globin genes are found in hundreds of millions of people. Yet, interest in hemoglobin has not been based entirely on its clinical importance or an intrinsic interest in the molecule. Instead, scientists often gravitated toward hemoglobin because it is easy to isolate. Simply draw a tube of blood, separate the red blood cells from plasma, and force them to burst open by suspending them in a dilute solution of sodium phosphate. The result is 99 percent pure solution of hemoglobin, with more where that came from anytime you need it. Protein scientists coming to work in the morning knew that they could get enough hemoglobin for the day's experiment before finishing their morning coffee, without searching for volunteers or sacrificing animals. There were many times in my past life as a hematology researcher when I would stick a needle into a vein in my own arm and draw blood if I, or one of my colleagues, needed hemoglobin, platelets, fresh whole red blood cells, or white blood cells. After a janitor walked into my lab early one morning and saw me with a tourniquet around my arm, stabbing one of my veins with a needle and syringe, I learned to do this in the privacy of a locked office. My red blood cells were very supportive targets for the *Plasmodium* strain used by a colleague of mine in her malaria culture system; and she frequently needed fresh red blood cells to feed her "babies," which I was happy to supply with a self administered blood draw. Hopefully, if I ever contract malaria, the parasites transmitted by mosquitoes will not find my red blood cells as appetizing as the cultured parasites did.

Before recombinant DNA made it possible to generate an unlimited supply of almost any protein (see chapter 14), researchers who worked on proteins that are present in vanishingly low concentrations, such as pituitary hormones, were not as lucky as hemoglobin researchers. They had to obtain large quantities of tissue–from thousands of pituitary glands harvested from slaughterhouse animals, for example–and conduct an elaborate purification procedure in order to get a usable concentration of their favorite protein. Any misstep along the way– a lab accident or an inadvertent sneeze–and you could lose the few precious granules of purified protein that took months to isolate.

Human genetics researchers are also attracted to the convenience of blood as a source of research material. The white blood cells present in a tube of blood provide an accessible source of human DNA, which can be isolated in a couple of hours using very simple purification methods.

The cloning and DNA analysis of globin genes culminated in the identification of the hundreds of different mutations responsible for inherited disorders of hemoglobin. Many human geneticists who received their scientific training during the late 1970s and early '80s began their careers working on the globin genes. For a couple of years, globin gene research was the only game in town.

The mutations found in the globin genes either alter hemoglobin's amino acid sequence, thereby affecting its chemical properties and its function, or reduce the efficiency at which the globin genes are expressed, causing a decrease in hemoglobin production. Some globin gene variants cause unusual but benign medical curiosities, such as a bluish discoloration of the blood; some cause severe illness that can handicap a person for life; and, some are downright deadly. Among the most common genetic disorders in the world are two hemoglobin disorders sickle-cell anemia (SCA) and thalassemia.

SCA is the most common recessive genetic disorder in black Africans and their descendants. A recessive disorder is one in which both copies of a gene are defective. In the case of SCA, the abnormality is a single mutation in a member of the globin gene family, the beta globin gene, located on chromosome 11. The mutation causes a change at amino acid number 6 of the 141 amino acids that comprise a single molecule of beta globin protein; the amino acid valine is added instead of glutamic acid. Depending on whether one or two copies of the mutation are inherited, this apparently simple change, a single amino acid substitution occurring in one of tens of thousand other proteins produced in the body, alters the chemistry of hemoglobin in a way that can have remarkable effects on a person's life, from protecting against cerebral malaria to destroying vital organs of the body.

Having two copies of each gene (except for the genes on the X and Y chromosomes in men which are present only as single copies) creates a certain degree of redundancy, since a single functional copy is usually sufficient. This is certainly true for hemoglobin. As long as one beta globin gene is functioning properly, the other can contain a serious abnormality, such as a sickle-cell mutation, without causing any adverse health effects. This is why carriers of recessive disorders, who harbor a single copy of a defective gene, are not sick. On the contrary, having one normal beta globin gene and one sickle-cell gene—a condition known as sickle-cell trait (SCT), which occurs in tens of millions of people in the world—is highly desirable, because it substantially reduces mortality from *falciparum* malaria. In some parts of Africa, the mortality rate of children infected with *Plasmodium falciparum* approaches 50 percent, but it is only about 10 percent in children with SCT, an 80 percent decrease.

We know from genetic studies that the sickle-cell mutation arose by chance in five individuals (four Africans and an Asian) who lived several thousand years ago in malarious regions of the world. Every person in the world with SCT and SCA is related to one or two of these five progenitors. The mutation has been so successful in malaria-endemic areas that between 10 percent and 20 percent of the populations there, and in some localities 30 percent, have SCT. The highest frequencies occur in pockets of African and among the Chetty and Kurmar peoples in India.

SCT is also found in Caucasians, although the frequency is much lower than in blacks. In some areas of northern Greece, southern Italy, Sicily, and southeastern Turkey, about 3 percent of the inhabitants have inherited the mutation. The sickle-cell gene in these regions appears to be of African origin, even though affected individuals are Caucasian, and was probably spread by traders, pirates, and North African slaves. The sickle-cell gene is also found throughout the Middle East, occurring in Saudi Arabia, Iraq, Iran, Syria, and Jordan, and among Palestinian Arabs. The "Arab" sickle-cell gene is derived from the Asian mutation, except in the southwestern regions of Saudi Arabia, where there has been an influx of the African variety.

Because of the admixture with Caucasian beta globin genes, the frequency of SCT is approximately 8 percent among African-Americans, a bit lower than the average rate among native black Africans. This reflects the 10 to 15 percent influx of Caucasian genes into the genomes of the descendants of African slaves. Nearly two million people in the United States have SCT. The gene is also fairly common among Hispanic Americans, many of whose ancestors acquired it in the slave trade that brought Africans to the Caribbean region. Based on the frequency of SCT in the United States, its occurrence in both members of African-American couples is approximately 1 in 150, which presents them with a 1-in-4 risk of conceiving a child who has two copies of the sickle-cell mutation, which causes SCA. These unlucky matches are sufficiently common to cause 72,000 people in the United States and more than 2 million people in the world to be born with SCA.

Although SCA is a very common genetic disorder, it was not recognized as a distinct disease entity by Western doctors until 1910 when Dr. James Herrick examined a smear of blood from a West Indian dental student who suffered from severe anemia, jaundice, shortness of breath and bone pain, all signs severe sickle cell disease. Herrick noticed that some of the patient's red blood cells were a peculiar shape resembling a sickle, a farm instrument used in harvesting grains.

Over the next decade, doctors continued to identify other patients with the same symptoms. Other than recognizing that the majority of patients shared an African heritage, and had the same peculiar blood cells, doctors were perplexed.

Sometimes important discoveries in science arise from the most mundane observations. This was certainly the case with the sickling of red blood cells, which E. Vernon Hahn and Elizabeth Gillespie, doctors at the University of Indiana Medical School, found in 1927 to be dependent on oxygen. They observed that when blood taken from an SCA patient was allowed to settle by gravity in a test tube, there were more sickled cells at the bottom than on top. When the cells were agitated, and thus mixed with air, sickled cells returned to a normal shape. They reasoned that, since the cells at the bottom of the tube were more oxygen-starved, sickling was caused by low oxygen levels. Their conclusion was established experimentally by removing or adding oxygen to a chamber containing SCA red blood cells, causing them to sickle and unsickle, respectively. We now know that there is a limit to the reversibility of the process; repeated cycles of sickling and unsickling will damage the cell's membranes sufficiently to make its sickled shape permanent.

Despite this knowledge, the underlying cause of sickling and its dependence on oxygen remained a puzzle for two more decades.

The solution began with a simple but now historic experiment performed in the late 1940s by Linus Pauling, widely considered the greatest biochemist of the 20th century. Pauling, who was a professor of chemistry at the California Institute of Technology, is probably known to most readers as an advocate–or a zealot, according to some–of megadose vitamin C for preventing colds and cancer, an unconventional idea that he embraced toward the end of his illustrious career. His main scientific contributions were discovering the hydrogen bond (the chemical bond that helps govern the structure of the organic molecules that make up life) and figuring out the three-dimensional configuration assumed by most proteins, which he dubbed the alpha helix, an achievement that garnered him a Nobel Prize in Chemistry in 1954. He came up with the idea for the alpha helix by spending a few hours drawing atoms on pieces of paper and folding them to simulate possible interactions between them, beginning his paper modeling while lying in bed recovering from a very bad sinus infection.

Pauling's interest in SCA came from a chance encounter in 1945 with the physiologist, William Castle, who told Pauling about the relationship between blood oxygenation and sickling. At the time, the prevailing theory maintained that sickling was due to a defect in the membrane of red blood cells. After his conversation with Castle, Pauling had the insight to consider the possibility that an abnormality in hemoglobin itself could be the underlying problem.

Pauling proved this by applying the newly discovered gel electrophoresis technique, a now-routine experimental tool that enables researchers to separate molecules that differ in size or electric charge, by passing them through a semisolid matrix (usually starch, agarose or acrylamide) placed in an electric field. Using this technique, Pauling observed that hemoglobin obtained from a patient with SCA migrated differently through a gel than normal hemoglobin would; it did this, he reasoned, because their amino-acid sequences differed. This was the first time anyone had ever detected a qualitative abnormality in a protein and linked it to a medical condition, endowing SCA with the distinction of being the very first "molecular disease."

Although Pauling spent most of his life working on protein chemistry, he also dabbled in DNA. A few years after his SCA discovery, Pauling used his knowledge of chemical bonds and proteins to develop a model for the structure of DNA. At the same time, James Watson and Francis Crick were building their own model based on the known chemistry of nucleotides, utilizing balls and sticks to simulate the carbon, hydrogen, oxygen, nitrogen, and phosphorus atoms that make up DNA, and the chemical bonds between them. They were aided by the now-legendary X-ray photograph of DNA, photo 51, taken by Rosalind Franklin, a chemist at King's College, Cambridge. Watson and Crick were surreptitiously shown photo 51 by their collaborator Maurice Wilkins, a colleague of Franklin's. Wilkins, Watson, and Crick were all on poor terms with Franklin, who had previously denied Watson and Crick access to her data. The X-ray proved to be the final piece of information that Watson and Crick needed to arrive at their successful solution to the structure of DNA.

Pauling had declared his interest in solving the puzzle of determining the structure of DNA. He was appropriately regarded by Watson and Crick as their most feared competitor. After all, the race pitted the greatest biochemist in the world (Pauling) against a physicist working as a biologist who was in his thirties and still trying to get his Ph.D. (Crick) and an arrogant young American who was obsessed with DNA but was supposed to be working on the structure of the muscle protein myoglobin (Watson). In fact, the pair was working on DNA behind the backs of their superiors, who had specifically ordered them to stick to their assigned projects, and warned them that the DNA project belonged to Wilkins and Franklin.

In 1952, a year before Watson and Crick reported their double-helical model of DNA, Pauling planned to attend a meeting of the Royal Society in London, where he hoped to catch a glimpse of the DNA X-ray photos taken by Franklin and Wilkins. Earlier, he had written to Wilkins' superior requesting copies of the images, but was refused. By appealing in person, Pauling hoped for a change of heart. However, because of his left-wing political views, Pauling never made the trip, as the government believed that it "would not be in the best interests of the United States" to grant him the freedom to travel. His passport was confiscated by authorities in New York when he tried to board a plane bound for England. The incident provoked so much criticism that he was permitted to travel to France and England later that year (although he still did not gain access to Franklin's photos) and to Europe the following summer, where he was able to see Watson and Crick's DNA model. His travel privileges were revoked again after he spoke out, shortly after that trip, in defense of J. Robert Oppenheimer, the embattled former director of the Manhattan Project.

Pauling was a pacifist with long a history of left-wing political views and activities, joining organizations (including the Emergency Committee of Atomic Scientists, headed by Albert Einstein, in 1946), and protesting against the loyalty oath required of federal employees, instituted by President Harry Truman in 1947 to keep suspected Communists and other subversives out of the government–a strategy zealously pursued by Senator Joseph McCarthy in the early 1950s. During World War II, Pauling had refused to work on the Manhattan Project. Instead, he worked on rocket propellants and artificial serum, for which President Truman awarded him a Presidential Medal of Merit in 1948. After the war, he was a proponent of controlling nuclear weapons, giving lectures, writing articles, and signing petitions, activities that brought him to the attention of the F.B.I. as a potential "security risk." He was denounced as a Communist by an F.B.I. informer in 1950, and was subpoenaed in 1951 to testify before the House Un-American Activities Committee, which used tactics of intimidation against allegedly subversive organizations and individuals. Pauling responded to the summons by issuing a public declaration that until 1935 he was "a registered Republican" and after that was a Roosevelt Democrat, and that he was not and had never been a member of the Communist party; however, out of principle, he refused to take a loyalty oath that was not genuinely relevant to national security. When Pauling was informed that he would receive the 1954 Nobel Prize in Chemistry (for his "research into the nature of the chemical bond and its application to the elucidation of the structure of complex substances"), he was afraid that he would be unable to get a passport to travel to Sweden for the award, due to his continuing troubles with the State Department; anti-Communist zealots in the U.S. government were even trying to get him ousted from his academic appointment at Cal Tech, where he was now head of the chemistry department. However, two weeks before the awards ceremony, with the support of Albert Einstein and other notables, the State Department granted him permission to travel and thus narrowly avoided a major public-relations debacle.

Without the benefit of the best available DNA X-rays, Pauling tried to generate a DNA model based on his knowledge of chemical bonds, but made a few fundamental errors in nucleic-acid chemistry and developed a convoluted triple-helical structure with the nucleotides pointed in the wrong direction. Watson and Crick were relieved when they saw the model; they knew it was wrong because they had come up with a similar one earlier and had realized that, chemically, it was not stable enough–the pieces did not quite fit. Knowing that Pauling would soon become aware of his error, they rushed to come up with something more suitable, which they did within a few weeks. Their correct determination of the double-helical structure of DNA and the A:T, G:C base-pairing rule became the most important biological discovery of the 20th century, leading to an understanding of how DNA replicates during cell division, as well as the coding system used by DNA to transfer information to mRNA and proteins. The double helix explained the chemistry of life; the Davids had beaten the biochemical Goliath. As Pauling put it, he lost the race to DNA to an "adolescent postdoc and an elderly graduate student."

Pauling's pacifism may have cost him a shot at the most important achievement in 20th-century biology, and a second Nobel Prize in Chemistry, but it did lead to a Nobel Peace Prize. He continued his anti-war activism and was a leader in the fight against atmospheric testing of nuclear weapons, sponsoring a petition signed by more than 10,000 scientists opposing nuclear testing and proliferation. When he was awarded the 1962 Nobel Peace Prize, he became the only person ever to win two Nobel Prizes solo (most awards are shared by one or two others). His peace prize was severely criticized in the United States, and *Life* magazine went so far as to call the award the "Weird Insult from Norway."

Although Pauling had shown that SCA was caused by an abnormal hemoglobin, he did not pinpoint the exact nature of the problem. That was left to Vernon Ingram, a member of the Cavendish Laboratory in Cambridge, England, which also included Crick and several other notable figures, many of whom went on to win Nobel Prizes. Using new protein analytical techniques, one of which was the ability to determine the amino acid sequence of proteins, a brilliant invention conceived by Fred Sanger, a Cavendish scientist who later went on to develop a technique for determining the sequence of DNA (both of which resulted in Nobel Prizes), Ingram established that sickle-cell hemoglobin differed from normal hemoglobin by a single amino acid at position six.[10]

---

[10]Other notable members of the Cavendish lab included Max Perutz and John Kendrew, who won a Nobel Prize in 1962 for being the first to determine the three-dimensional structure of a protein, and Sydney Brenner, another Nobel laureate, who, along with Crick and French biologist and Nobel laureate Françoise Jacob, helped conceive the idea of a messenger RNA intermediary that carried the genetic code from DNA to the cell's protein-synthesis apparatus. Brenner also introduced a 1-millimeter-long nematode called *Caenorhabditis elegans* to the research community; an organism which is now one of the most commonly used tools for geneticists (the hardy worms were the only survivors of the 2003 shuttle disaster–they were on board to study the effect of weightlessness on development and were found alive in a cannister that remained intact after the shuttle burn-up and crash).

The entire clinical spectrum of SCA can be traced to a unique property of sickle-hemoglobin not found in normal hemoglobin: it attaches to itself over and over again, building up to form a large pipe-like fiber. Sickle-hemoglobin fibers damage the membrane of the red blood cells, causing the characteristic sickle shape. Picture a handful of swizzle sticks placed in a sealed flexible bag that is smaller in length than the sticks are, making the sticks bend a little in the middle and changing the shape of the bag to accommodate them. The attraction that sickle-hemoglobin molecules have for each other accelerates under low oxygen conditions, explaining why the number of sickled cells increases when oxygen levels decrease.

Two general problems are encountered by red blood cells stuffed with sickle-hemoglobin fibers. First, the cells are destroyed faster than normal. The surveillance system that normally removes old, damaged red blood cells after they have reached the end of their 120 days of life destroys the damaged, misshapen sickled cells many weeks before their scheduled demise. Although the bone marrow tries to compensate by generating new red blood cells at an accelerated clip, the rate of destruction exceeds production, causing anemia.

The second problem is the tendency for sickled red blood cells to "stick" and obstruct blood flow. Normal red blood cells are shaped like a disk that is thicker on the outer rim than at the middle, similar to a tire or a bagel without the hole, and have a "spongy" consistency that makes them pliable and bouncy, facilitating their movement through very small blood vessels, the "microcirculation." The protein scaffold that makes up the red blood cell "cytoskeleton" gives cells their pliability, similar to the construction design of bridges that enables them to sway in strong winds and during earthquakes without being damaged. Sickled cells are not as pliable as normal red blood cells, reducing flow through the microcirculation. In addition, sickled cells tend to stick to the walls of small blood vessels. Like a series of speed bumps, progress through the microcirculation slows when sickled cells attach to vessel walls. Stagnant blood flow prevents normal oxygenation, which damages tissue, causing the constellation of symptoms seen in SCA.

The hip and knee joints, the kidneys, skin, liver, lungs, spleen, and brain are parts of the body commonly damaged by SCA, but nearly any organ can be affected. Damage to the spleen causes an increased vulnerability to certain infections, especially *Streptococcus pneumoniae*, the most common cause of bacterial pneumonia. An untreated patient can die within hours of developing the first sign of a severe infection; I have seen sickle-cell patients hospitalized for infection, who, although they were receiving intravenous antibiotics and appeared to be stable during morning rounds, were dead by the end of the day, victims of the overwhelming type of bacterial infection we call sepsis.

Chronic obstruction of blood flow can destroy the hip and knee joints, causing crippling pain and loss of mobility, resulting in artificial joint replacement. S.C.A often affects the skin in the lower extremities which ulcerates when deprived of adequate blood flow. Skin ulcers can extend into the subcutaneous tissue down to bone, requiring transfusion therapy, surgical excision, and skin grafts. When SCA affects small blood vessels in the brain, it can cause strokes that no different from those suffered by elderly people with cardiovascular disease. Even the penis cannot escape the effects of sickled red blood cells. Normal erections occur when veins in the penis fill with blood. These veins are vulnerable to the occluding action of sickled red blood cells. When this occurs, blood fills the penis but does not empty from it, causing priapism, a prolonged painful erection (priapism is named after Priapus, the Greek god of fertility).The painful erection can go on for hours, even days; sometimes the backed up blood has to be aspirated with a needle inserted into the penis.

The most common manifestation of SCA is a sudden obstruction in blood flow that leads to the acute onset of severe pain known as the "painful crisis." This very common complication is the major reason that patients who have SCA seek emergency care. The pain can occur in bones, abdomen, or chest, and may last from hours to days. Female patients with SCA report that some attacks of painful crisis are worse than labor pain. Other patients have described the pain as a "total body toothache" that won't go away. The pain is severe enough to require narcotic pain killers such as Demerol or morphine for relief.

SCA patients' need for potent narcotics is a frequent cause of conflict: the battle over obtaining adequate pain relief versus the doctor's fear of causing an addiction or being duped by a drug trafficker. This is not an issue that comes up only in SCA; any chronic pain problem requiring narcotics, fairly or unfairly, raises a red flag. But only SCA brings racial factors into the picture, since most patients are Blacks and Hispanics, whereas most doctors are not. Every SCA patient with a painful crisis has, on some occasions, experienced the humiliation of having to beg for narcotics from doctors and nurses who did not appreciate the pain they were experiencing. Not surprisingly though, on occasion, a serious drug problem does surface in a patient with a chronic pain condition. I had one patient with very severe SCA who managed to beat the system for a number of years. He was more than six-feet tall but weighed only 110 pounds. By the time he was thirty, he had already received two artificial joints (a hip and a knee) and was ready for a third. He looked and talked like an emaciated version of Muhammad Ali. Because of his damaged joints, he walked like the Tin Man from the Wizard of Oz (he and I used to joke about having his joints oiled). I gave him prescriptions for oral narcotics that he could take at home for painful crises to help keep him out of the emergency room and out of the hospital, where he was spending about 10 percent of his adult life as an in-patient recovering from suspected sickle-cell crisis attacks. Despite having a very bad case of SCA, he had developed a reputation as a drug-seeking manipulator in several emergency rooms in New York City. Our relationship came to an abrupt end after five years. A homicide detective visited my office and informed me that my patient had been murdered. He was apparently a major peddler of prescription narcotic pain killers and was bludgeoned to death in his apartment during a drug sale or robbery. The police found dozens of current and old prescriptions for narcotic painkillers that I had written, as well as prescriptions from a number of other doctors in the area. The detective showed me the crime-scene photographs. The head wound that killed him was just at the site where he had developed a bald spot from lying in the same position with his head resting on the edge of a sofa for so many hours a day, immobilized because of joint pain, spending his time watching television, and, apparently, talking on the phone making drug deals.

Like SCA, the large group of inherited hemoglobin disorders known collectively as thalassemia also evolved because of Darwinian selection against falciparum malaria. While SCA is caused by a mutation that leads to a qualitative change in globin protein–an alteration in an amino acid on the beta globin gene–thalassemias are caused by mutations that lead to quantitative changes: the hemoglobin molecule itself is normal, but the amount produced is too low. This happens when mutations disrupt normal gene expression in some manner. Since 1979, when the globin genes were first isolated, a steady stream of genetics research has resulted in the discovery of hundreds of different thalassemia mutations, all of which have the same effect on red blood cells: instead of displaying the normal robustness due to a full package of hemoglobin, they are weak-looking, thinned-out and small, because of a shortage in the amount of hemoglobin they produce.

The name *thalassemia* is derived from the Greek word *thalassa* (Θαλασσα for Greek scholars, and fraternity enthusiasts) meaning "the sea", to denote the high frequency of the condition in people who come from the Mediterranean region. Like SCA, thalassemia is an autosomal recessive disorder. Inheriting only a single thalassemia abnormality, which occurs in one form or another in about 10 percent of Italians and Greeks, 15-20 percent of people from Southeast Asia, and 30 percent of African Americans, leads to a benign condition known as thalassemia trait (or thalassemia minor). Like sickle cell trait, thalassemia minor also protects against falciparum malaria. However, two parents with thalassemia minor have a 1-in-4 chance of conceiving a child who has two copies of the mutation that codes for the condition, which leads to the fatal disorder, thalassemia major. Because of differences in the type of thalassemia mutations found in different ethnic groups, thalassemia major is primarily, but not exclusively, found in the Italians and Greeks. Patients with thalassemia major can be saved by monthly transfusions of blood, which have to be taken for the duration of their lives. However, by the time thalassemia patients reach the age of 30 or so, the liver and heart are overloaded with iron, which comes primarily from the hundreds of units of transfused blood they receive over the years, leading to premature death.[11] The idea that high frequency of the sickle-cell mutation could be related to a survival advantage against malaria was proposed in the late 1940s and early 1950s by J. B. S. Haldane and other geneticists. In some parts of Africa, the mortality rate of children infected with *Plasmodium falciparum* approaches 50 percent, but it is only about 10 percent in children with SCT, an 80 percent decrease. Even though SCA is usually fatal among people in underdeveloped countries, in parts of the world where SCT occurs at a 10 percent frequency, only 1 in 400 children will be born wth SCA and die because of it. When an 80 percent reduction in mortality from a widespread infectious disease such as malaria is provided by a trait found in 10 percent of the population that is itself responsible for the death of only 1 in every 400 children (caused by the bad luck to inheriting two copies of the protective gene), that is considered an acceptable trade-off by evolutionary standards. Under such severe selective pressure, rapid Darwinian selection for a protective gene could occur in just a few hundred years.

The biological mechanism for the protective effect of SCT has been difficult to establish, primarily because cultures of *Plasmodium falciparum* in laboratory incubators and flasks do not mimic the complex events that take place in the body when red blood cells are infected with the parasite. One possible explanation is that infection with *Plasmodium falciparum* causes the red blood cells to use up oxygen, which induces them to stickle; even though red blood cells do not use oxygen for energy production (see chapter 4), the parasites living inside do. Sickled cells are then removed from the blood by phagocytic cells in the spleen. Parasites in the trapped cells are destroyed before they have a chance to mature and spread.

---

[11]Every molecule of hemoglobin has four atoms of iron-when hemoglobin is metabolized in the body, which occurs whenever a red blood cell dies, its iron is recovered and stored in the liver and bone marrow. Iron tends to accumulate to toxic levels in patients who have received more than a hundred units of blood.

The lives of patients who have SCA have improved substantially during the past 40 years because of the availability of blood transfusion therapy to treat life-threatening complications, which occur a few times in almost every SCA patient, and the use of potent antibiotics to treat serious infections. Now, many patients in the United States live into their sixties and beyond. In Africa, most children with SCA die in childhood because these treatments are not readily available and because of the general poor health of children in underdeveloped countries. But blood transfusions and antibiotics are non-specific therapies–they are by-products of modern medicine. The expectation from all of the research done on SCA over the decades has been that a more definitive cure, specific to SCA, would be developed. This has not happened yet, although there are some promising new treatments being tried, and others are on the way, including gene therapy.

The one area where research on the genetics of globin genes has really paid off–and the most controversial–is prenatal diagnosis. DNA obtained from the fetus can now be used to diagnose any life-threatening hemoglobin disorder. The most accessible sources of fetal DNA are from cells recovered during amniocentesis and from a portion of the placenta called the chorionic villus.

Once fetal DNA has been obtained, a genetic diagnosis can be made very rapidly using an application of the Polymerase Chain Reaction (PCR), the DNA amplification procedure that has revolutionized many aspects of human genetics and medical diagnosis. PCR-based analytical techniques are very fast and very sensitive, and a diagnosis of SCA or thalassemia major in a fetus can be made in only a few hours using a relatively small quantity of DNA.

For certain disorders caused by a single, well-characterized mutation, such as SCA, it is possible to push the limits of PCR's analytical sensitivity by testing only the minuscule amount of DNA found in a single cell. The technique can be performed on a cell derived from an eight-cell embryo prepared by *in vitro* fertilization. Only embryos without the mutations are transferred into the mother's uterus; children born from a seven-cell embryo are indistinguishable from those who develop from the regular eight-cell stage.

Prenatal testing for SCA is performed less frequently than for thalassemia. This is largely due to the view in the United States and Europe that SCA is not necessarily a fatal illness. But there are other factors, one of which is racial politics. A prevailing level of mistrust exists among people in the African-American community regarding medical research, and this limits their participation in trials of new techniques and their acceptance of new genetic technologies. The notorious syphilis experiments by the U.S. Public Health Service (USPHS), involving researchers from the Tuskegee Institute, set the stage for the negative atmosphere. From 1932 to 1972, the USPHS conducted a study on 600 African-American sharecroppers in Alabama, to follow and document the natural history of untreated syphilis. The men in the study who had syphilis were not told that they had the disease (only that they were being treated for "bad blood"), and were not provided with treatment when it became available (penicillin was introduced as a treatment for syphilis in the 1940s).

Even before prenatal diagnosis became available, widespread screening for SCT to help identify couples at risk for conceiving a child with SCA was controversial when it was introduced in the 1960s. SCT in children and adults can be easily diagnosed by a simple blood test. However, the idea of universal screening was criticized as black genocide by some, since the only options offered by genetic counseling available at the time for couples who wanted to have children and were both carriers of SCT was either considering not having children or choosing different partners. In a letter to a friend written in 1966, even Linus Pauling suggested that... "the time might come in the future when information about heterozygosity in such serious genes as the sickle cell anemia gene would be tattooed on the forehead of the carriers, so that young men and women would at once be warned not to fall in love with each other."[12] In addition, it was feared that a diagnosis of SCT would lead to job discrimination and problems obtaining life insurance, realistic concerns at the time. Although there are no serious medical consequences to having SCT, insurance companies charged higher premiums. African-American

---

[12]http://osulibrary.oregonstate.edu/specialcollections/coll/pauling/blood/quotes/all.html

women with SCT who wanted to work as stewardesses were discriminated against by airlines, who refused to hire them because of the fear that repeated high-altitude flights would cause their red blood cells to become sickled. In fact, sickling can sometimes be provoked in those who have SCT under the low-oxygen conditions found at very high altitudes, but the air used on commercial airlines has enough oxygen to prevent this. A person with SCT would have to be exposed to air at an elevation as high as Mount Everest to provoke sickling. Men with SCT were not permitted to join the Air Force for similar reasons.

Consequently, prenatal testing for SCA in the modern DNA era has not been utilized to its fullest potential. However, the idea of prenatal testing is more acceptable in African countries that have acquired the expertise to perform the analysis, such as Nigeria. This could reflect the greater urgency of a SCA diagnosis in Africa, where most patients die young, and the absence of racial barriers in most African countries, where most of the doctors and patients are black.

Racial politics has clouded other issues related to SCA. During the late 1960s and early '70s, there were arguments in the African-American community—well founded at the time—that funding for SCA research was too low, being a fraction of the funding allotted for studying cystic fibrosis, the most common autosomal recessive disorder in white people. In 1971, President Richard Nixon called for increased funding for SCA research, resulting in the passage of the National Sickle-Cell Anemia Control Act and the appropriation of $115 million for research. However, this was criticized by some as pandering to African-American voters for the upcoming 1972 election (the election that precipitated the Watergate break-in scandal). There was also criticism that the money would be better spent on social services such as housing and education. In addition, some members of the medical community argued that more research funding should be applied to hypertension and diabetes, which disproportionally affect African-Americans, and have a far greater impact on health than SCA.

In some parts of the world where the prevalence of inherited hemoglobin disorders is very high, prenatal diagnosis is a public health issue, not simply a private one. There has been a 90 percent decrease in the number of babies born with thalassemia major in Sardinia, for example, as a result of an aggressive screening campaign. Without prenatal screening, the island would use up its entire health-care budget treating children born with thalassemia major to sustain them during their short, damaged lives.[13] Presumably, the overwhelming majority of families at risk who aborted fetuses with thalassemia major later conceived children free of the disorder.

What is most remarkable about SCA and thalassemia is the powerful clinical effect caused by such a small change in the genome. Both are due to single nucleotide changes occurring in a genome made up of three billion nucleotides, a chemical difference equivalent to a single letter change in a million-page manuscript. Yet this seemingly simple change has had a significant impact on human health. Inheritance of a single sickle-cell or thalassemia mutation has successfully infiltrated the human population by reducing mortality from *Plasmodium falciparum* infection. The mutations, which arose by chance in a handful of individuals at the dawn of civilization, are now found in hundreds of millions of people throughout the world. The price of success has come at the expense of the unlucky millions who inherit two copies of the protective genes, a condition that initiates a chain of molecular events resulting in debilitating and potentially fatal hemoglobin disorders, with symptoms as diverse as priapism, the destruction of joints, and stroke. But having two copies of harmful beta globin genes has not merely brought misery to families over the past few thousand years, it has also created moral dilemmas–whether or not to undergo prenatal diagnosis and pre-implantation selection–and has led to scenarios that brought racial discord to the surface. SCA, the genetic illness that has been called the first molecular disease, may also qualify as the first with a social agenda.

---

[13]Aggressive screening has also been very effective in reducing the incidence of Tay-Sacks disease, a fatal autosomal recessive disorder found primarily in Eastern European Jews. Identifying families at risk and carrying out early prenatal diagnosis has essentially eliminated the disease from North America.

CHAPTER 7

Tainted Water

A host of different bacteria, viruses, and parasites have prospered because of their predilection for the human gastrointestinal tract and the surprising ease with which their progeny can be transmitted from the feces of one person to the mouth of another. The upshot of this infectious cycle is acute diarrhea and chronic parasitic disease. On a grander scale, the "fecal-oral" route of infectious disease has been so pervasive in human society, and has caused so many fatalities, especially in children under the age of five, that it has provided the engine for numerous changes to occur in the human genome, one of which is the gene responsible for the inherited disorder cystic fibrosis.

Many animals have an innate sense of the health risk from feces, adopting complex behaviors to bury their waste or separate it in some manner from their living space. These behaviors are programmed traits that probably evolved by conferring a survival advantage against deadly infectious diseases transmitted through body waste. Yet, for the most intelligent creatures on Earth, humans have been curiously ignorant about the fecal transmission of disease. Throughout most of human history, fecal waste has been treated carelessly, dumped too close to living quarters, or collected in cesspools that were allowed to meander into drinking water. It was not until the 19th century that the impact of human waste as a source of disease was fully appreciated and washing hands after defecating was advised.

During the 19th century, doctors were as irresponsible as laymen about washing their hands. One of the more pervasive infectious-disease problems of the mid-19th century, before the work of Louis Pasteur, Robert Koch, and Joseph Lister established the germ theory of disease and proved the value of antisepsis, was the high rate of infection after childbirth (puerperal sepsis, childbed fever). This was exacerbated by doctors going from an autopsy or dissection table directly to a birthing room without washing up. This practice increased an expectant mother's chance of dying if she chose a medical man over a midwife to deliver her babies. By the mid-19th century, one of every six new mothers who gave birth in maternity hospitals in Paris died of infection. Although puerperal sepsis is usually caused by the skin contaminants *Streptococcus* and *Staphylococcus,* which are not transmitted by the fecal-oral route, it illustrates the mind-set of nineteenth-century doctors regarding basic hygiene. A Hungarian doctor, Ignac Semmelweis, eventually made the connection between the high rate of puerperal sepsis in Austrian hospitals and contaminated hands. Instead of being praised and rewarded for this insight, Semmelweis was persecuted by his Austrian colleagues, who refused to accept their part in the deaths of new mothers. Semmelweis's cause was not helped by his tactless approach to the problem: he accused some of the most established physicians in Austria of murder. He eventually suffered a mental breakdown and ended his days in an asylum.

Semmelweis's observations were filled with excellent epidemiological intuitions, but he missed an opportunity for greatness when he proposed that puerperal sepsis was caused by "cadaver particles," a vague and unsatisfying hypothesis. Had he taken the trouble to look at pus from contaminated patients under a microscope—a now-obvious first step that any novice biology student would take today—he might have seen the teeming masses of cocci bacteria responsible for puerperal sepsis, which would have launched the age of bacteriology years before Pasteur and Koch made their discoveries.

Although cats instinctively bury their waste, fecal attentiveness in humans is more a learned trait, acquired through constant badgering. Telling children to wash their hands after going to the bathroom rivals reminders about brushing their teeth and limiting the amount of sweets they eat as parents' most common admonitions. Yet, washing hands after defecating and after changing diapers is not universally practiced, as is evident from various surveys and direct observation, as well as from the extraordinarily high rates of intestinal parasites that exist in much of the world, which are often transmitted by fecal matter going from hand to mouth. Various surveys carried out on men and women using public washrooms by "bathroom spies" hired by the American Society of Microbiology showed that approximately one out of three men and one out of five women do not wash their hands after using the bathroom. In the survey, men at Chicago's O'Hare airport, New York Cities Grand Central Station and Times Square, and baseball fans at Turner Stadium, home of the Atlanta Braves, were the worst offenders. Men and women in Toronto had the highest rate of washing their hands, more than 95 percent, probably reflecting that city's encounter with the recently emerging disease of SARS at around the time the study was conducted, during which washing hands was being promoted as a public health measure. The studies were initiated after the organization became suspicious of the high rate (95 percent) reported by men and women when asked in a telephone survey about their practice of washing their hands. This turned out to be much higher than the rate that was found by spying on people in public bathrooms, giving new meaning to the term "dirty liar."

The importance of basic hygiene is apparent from studying the bacterial load needed to cause infection. Some bacterial pathogens that attack the gastrointestinal tract, such as *Shigella*, can initiate life-threatening infections with as few as one hundred organisms, which can easily be transmitted as an invisible microparticle on the hands of the unwashed.

In parts of the world with excellent sanitation, the failure of people to wash their hands adequately can really only be explained by sheer laziness. Even doctors and nurses, who should know better, need some gentle prodding. The practice of washing up between patients has to be drummed into their heads during one-hour seminars that they must attend every year or two in order to maintain a license to practice. Actually, though, the transmission of fecal contaminants by doctors and nurses is less of a problem than their transmission of skin bacteria from one patient to the next, especially in hospitals, which house some of the most pernicious, antibiotic-resistant microbes in existence. Despite the frequent seminars in infectious disease control, only a minority of doctors and nurses adhere to the practice of washing hands between their contacts with different patients, and their lax hygiene contributes to the tens of thousands of deaths in hospitals every year from sepsis.

Fecal contamination by medical workers occurs quite frequently in nurseries though, where the failure to wash one's hands properly can infect newborns with the intestinal bacteria *Escherichia coli* and *Enterococci* and thus cause deadly infections in newborn babies. One of the hiding places for microscopic fecal matter in nurseries is the inner concave fold of large artificial nails adorning the fingers of some nurses.

Untidy food handlers who touch food that is subsequently kept at room temperature for hours, allowing time for dangerous bacteria to grow or produce toxins, are a common source of contamination. Every year, about 9,000 deaths occur in the United States from food poisoning, out of nearly 100 million cases. Another source of fecal-oral contamination is undercooked or raw shellfish from fresh water that has been inadvertently contaminated by human waste, which usually occurs when under-treated waste is dumped too close to a fishing area. Despite the great expanse of Earth's oceans, contaminated bathroom waste dumped into open water by cruise vessels and other ships can sometimes find its way into the food chain.

Fecal spread of disease has a cultural basis in underdeveloped parts of the world. In some rural areas of Pakistan and India, it has been the custom to apply ghee, a clarified butter used in cooking, to umbilical wounds in newborns. When ghee is heated with animal dung, which is a major source of fuel in much of the world, it can become contaminated with *Clostridium tetani* (the bacterium that causes tetanus), a common intestinal inhabitant in domesticated farm animals. A more egregious cause of tetanus occurs in northern Pakistan when dried cow dung is applied directly to the skin of infants the way talcum powder is used in the West, after which the babies are wrapped in a tight sheepskin cover. Exposure of the raw umbilical stump to the powdered dung in the tight airless environment ureated by the sheepskin support establishes a beautiful niche for *Clostridium tetani,* since it is an anerobic bacterium that grows best when oxygen levels are low. Before local antibiotics became available in the last decade, nearly 4 percent of Punjabi newborns died of tetanus. In Kenya and Tanzania, infection is also common in Masai boys who have cow dung rubbed onto the raw wounds resulting from circumcision rituals performed when they reach puberty.

Humans are also victims of the fecal-oral transmission of bacteria that infect animals. *Salmonella* species that commonly cause food poisoning in humans are fecal bacteria that infect poultry, occasionally contaminating the chickens and eggs we consume. The poultry industry's practice of maintaining the maximum number of chickens confined in the smallest possible space has led directly to a modern explosion in the number of cases of *Salmonella*, and its indiscriminate use of antibiotics to keep the chickens disease-free has resulted in an increase in drug-resistant *Salmonella* strains.

Socio-economic factors are primarily responsible for the fecal-oral spread of disease in most of the world. According to the World Health Organization, 78 percent of people in under-developed countries lack a source of clean water, and 85 percent lacks a proper means of disposing of human waste. The problem has been exacerbated in poor countries by large-scale urban migration. In South Africa, where fresh water in poor communities has to be bought, a serious outbreak of cholera occurred among people who drank from contaminated sources because they could not afford to buy clean drinking water. However, simple, inexpensive solutions are sometimes possible. Recent studies by Dr. Stephen Luby of the CDC conducted on people living in very poor areas of Karachi, Pakistan, showed that promoting hand washing and providing free soap can reduce the mortality from infectious diarrhea by half.[14]

---

[14]Reported by Reuters news service June 3, 2004.

The principal consequence of the fecal assault on mankind is that almost every human on Earth is infected with one diarrhea-causing microbe every one or two years, and the rate is higher in children. By the time a child in the United States has reached the age of five, he or she has experienced approximately a dozen episodes of infectious diarrhea. About 9 percent of all childhood hospitalizations are for severe diarrhea. In addition, between 20 and 50 percent of foreign travelers will develop a case of infectious diarrhea from any of nearly a dozen different strains of bacteria, viruses, and parasites (the toxin producing enterotoxigenic *E. coli* is the most common).

Yet there is also a positive side to the less-than-pristine bathroom habits of human beings. In addition to becoming exposed to infectious agents by the fecal-oral route, we also ingest harmless bacteria that are important occupants of our intestinal tract (see chapter 1). Also, repeated exposure to a limited number of some infectious microbes, which may lead to a mild case of diarrhea, could provide immunity to future, more large-scale attacks by those organisms. A colleague of mine from India used to routinely serve her children unwashed fruit, while living in the United States, believing that it made their yearly trips to India easier on their gastrointestinal system. Moreover, exposure to certain gastrointestinal infections in childhood rather than in adulthood can sometimes be advantageous. One example is poliovirus, which is acquired through the fecal-oral route. Childhood exposure usually leads to a harmless gastrointestinal illness or causes no symptoms at all. Asymptomatic infections provide lifelong immunity to future poliovirus exposure. Problems only occur in a small minority of people when the virus leaves the gastrointestinal tract and travels to the spinal cord, where it causes permanent paralysis by damaging motor neurons, the nerve cells that control muscle movement. Although paralytic polio can occur in anyone regardless of age, it happens more commonly in adults who are exposed to poliovirus for the first time. So, if one is going to get exposed to poliovirus, it is much better to come into contact with it during childhood, the chances of which are higher in an unsanitary environment. The peak in the polio epidemic in the United States that occurred during the first half of the twentieth century coincided with improved sanitation conditions. Polio is one of the rare diseases where childhood squalor is somewhat advantageous. If Franklin Delano Roosevelt's mother had allowed young Franklin to play around in the dirt or cavort with a kid from a less fastidious home, he might have suffered from an additional short-lived gastrointestinal infection as a child and avoided paralytic polio as an adult.

The propensity for fecal-oral contact between humans has been exploited in the worldwide polio-eradication program. The most popular polio vaccine administered throughout the world is the Sabin oral vaccine, which is a live virus that has been weakened or attenuated in the laboratory. The virus almost never causes disease, but it is still capable of being transmitted by fecal-oral contamination. Administering the Sabin virus to a child often results in the vaccination of other members of the family and close contacts, by inadvertent fecal-oral transmission. Public-health officials refer to this as "herd immunity." Although not as effective as a series of scheduled vaccinations, person-to-person transmission of the Sabin virus is one of the reasons for the enormous decline in polio around the world, since people who have avoided vaccination can still get immunized that way. The capacity to deliver attenuated polio virus by herd immunity was one of the reasons the Sabin vaccine replaced the original polio vaccine developed by Jonas Salk, which is derived from a killed virus that is only effective in people who are directly immunized.[15]

Another benefit of fecal-oral contamination is its potential for preventing allergies and asthma. In the industrialized world over the last century, there has been a tremendous rise in the prevalence of asthma and allergies, disorders that rarely occur in most poor countries. One theory, the "hygiene hypothesis," attributes this to reduced microbial exposure in childhood.

---

[15]Because the Sabin vaccine is a live virus, it can cause infectious complications, including, very rarely, the development of polio in some people. This happens in about one in a million vaccinated people. During the polio epidemic in the early-to mid-20th century, which affected nearly 1 percent of the population, the risk of vaccination was greatly outweighed by the risk of contracting polio. However, now that polio has become extremely rare, the risk of vaccination is approaching the risk of contracting the disease. Consequently, there is now a movement to return to the Salk vaccine, which has fewer side effects.

Allergies and certain types of asthma are triggered when an allergen interacts with the IgE class of antibodies, causing the release of histamine and other active substances from several cell types, which are responsible for allergy symptoms and asthma attacks. Widespread exposure to bacteria in childhood increases the repertoire of antibodies from the IgA and IgG classes which, when present, may sequester allergens later in life, preventing or reducing their attachment to IgE.

In the United States, infectious diarrhea in an adult is usually a brief, self-limiting illness treated with over-the-counter medications and fluids. For a traveler to a foreign country, a case of diarrhea can ruin a trip, and is therefore viewed with a certain degree of trepidation during vacation planning. Loading up on supplies of the antibiotic Cipro (Ciproflaxin) and anti-diarrhea medication, and discussing foods to be avoided and the wisdom of using bottled water for brushing one's teeth have become as familiar a part of preparing for foreign travel as vaccinations and passports. However, in children, diarrhea can be deadly. The loss of fluid from watery diarrhea, exacerbated in toddlers and children by their small body size relative to the surface area of their intestines, can lead to severe dehydration and circulatory shock. Several hundred children die from diarrhea every year in the United States. In underdeveloped countries, diarrhea is a staggering problem that results in the death of about three million children every year, nearly ten thousand a day. It is the second leading cause of infant and childhood mortality in the world, after respiratory infections. According to the World Health Organization, infectious diarrhea is the leading cause of potential years of life lost.

The problem of diarrhea in underdeveloped countries today is similar to what it has been on all of humanity for most of our existence. Until the 20th century, severe diarrhea brought on by the bacterium that causes cholera and by other microbes has been a major cause of premature death, providing the engine for changes to occur in the human genome.

The loss of life from diarrhea in the modern world is one of the greatest tragedies confronting humanity; it is almost always completely treatable, and largely preventable. The infectious microbes themselves are often not toxic in the classic sense, with the exception of some strains of *E. coli, Salmonella, Shigella,* and a few other bacteria, which can cause either severe systemic infection or dysentery. *Vibrio cholerae,* the spiral bacterium that causes cholera, and Norwalk Virus, a common cause of infantile diarrhea, have no significant toxic effects on cells. The main problem caused by these organisms is their capacity to induce relentless watery diarrhea that can result in severe dehydration. If lost fluid is replaced, diarrhea usually runs its course, so to speak, and recovery is complete. Antibiotics are necessary to treat some gastrointestinal bacterial infections, but for cholera they are a luxury, helpful in shortening the course of illness and reducing the number of excreted bacteria, but not really essential for complete recovery. For viral diarrheas, antibiotics are useless, and even counterproductive, because they disturb the normal balance of bacteria that exists in the gastrointestinal tract.

Anti-diarrhea medications are often unnecessary and largely ineffective for severe watery diarrhea. For some gastrointestinal infections, anti-diarrhea medications are dangerous because they allow pathogenic bacteria to stay in the intestines for a longer period of time, thereby prolonging the duration of the illness. Simply preventing and treating dehydration is the best regimen. After a few days, *Vibrio cholerae* and other such organisms are overcome by the immune system. Although intravenous fluids are often needed to treat severe dehydration, especially if blood pressure has fallen, oral hydration is sufficient in the early phases. Simple oral hydration can reduce the cholera death rate from 50 percent to less than 0.5 percent, a hundred-fold reduction. A special blend of salts and sugar dissolved in clean water, called Oral Rehydration Therapy (ORT), has been developed specifically to help treat cholera and other infectious causes of watery diarrhea. The formula is based on the calculated salt and water loss in diarrhea, as well as on the physiological discovery that sugar inhibits salt and water loss from the small intestine. The constituents of this magic elixir, which has saved tens of millions of lives around the world, is very similar to the formulations of power drinks, or of Coke and Pepsi, but without the artificial dyes and bubbles. The drinks we casually consume after strenuous workouts in the gym and at fast-food restaurants are similar to the oral transfusions of life to a dehydrated child in Africa and Asia.

Because of its simplicity and low cost (which make it possible to use in underdeveloped countries) and its life-saving potential, ORT has been hailed as one of the most important medical discoveries of the twentieth century. Yet, despite the existence of this incredibly simple treatment, millions of children in the world still die every year from diarrhea. Lack of resources is primarily responsible for this unforgivable tragedy. Preparing ORT does require some elementary training, as well as filtered or sterilized water, but it's simple enough for uneducated people to learn. The challenge is getting the message to every one of the billions of people at risk.

Once severe dehydration leads to circulatory collapse, ORT loses its effectiveness. Then, intravenous fluid and salt replacement are needed. But, there are too few health facilities in poor countries capable of providing sterile intravenous fluids to treat all children with severe dehydration; by the time a victim in a small village can be transported to one, usually several hours away, it may too late.

Why is watery diarrhea so dangerous? The body's control over water and electrolytes is a tightly regulated affair primarily involving the hypothalamus (an autonomic regulatory center in the brain), pituitary gland, and kidneys; the adrenal glands, liver, heart, and gastrointestinal tract also play a role. These various organs maintain water and electrolyte levels in blood and tissue within the range needed for cells to properly function. Every cell in the body is responsible for balancing its own supply of water and electrolytes, which is achieved by a variety of ion pumps and ion channels through which sodium, calcium, potassium, chloride, magnesium, and bicarbonate flow. However, individual cells are unable to sustain normal water and electrolyte levels when total body reserves are low. As the blood and extracellular space lose salt and water, so do cells.

The human body is 60 percent water, and roughly 2 to 3 percent of that is lost and replaced every day–approximately one liter in the urine, a few hundred milliliters in sweat, another several hundred milliliters in feces, and a small amount of water in the process of breathing, since exhaled air contains more water vapor than inhaled air. Specialized cells called osmoreceptors located in the thirst center in the hypothalamus perceive water loss, and stimulate drinking behavior by sensing the increase in body salt concentration (mainly sodium chloride) caused by dehydration. The hypothalamus also instructs the pituitary gland to secrete ADH (antidiuretic hormone), which increases water reabsorption from the kidneys. With fresh water sources in such short supply in the world, ADH is perhaps the most environment-friendly hormone in existence. Without it, we would excrete a wasteful five to ten liters of urine every day, and correspondingly, our consumption of water would have to increase five or ten-fold. There would simply not be enough fresh water on the planet to sustain billions and billions of water guzzlers if animal life evolved without ADH.

Regulation of water intake is quite precise. The kidneys excrete any excess water consumed, and deficiencies are corrected over time by stimulating a continuing thirst for liquids. Some mentally ill patients overload the thirst center by drinking tens of liters of fluids a day when they are not thirsty, which can actually dilute the concentration of sodium chloride in the blood to dangerously low levels, leading to brain swelling and coma.

Dehydration develops when the loss of fluids exceeds their intake, which can occur under a variety of circumstances, most of which are fairly obvious. Infants, the elderly, and any indisposed individuals who depend on caretakers for access to fluids can become dehydrated if their normal everyday loss of water is not adequately replaced. This is exacerbated when fluid loss mounts. An immobilized elderly person who develops a fever on a hot day will lose more fluid than usual in sweat, which can bring on severe dehydration quite rapidly–a fairly common occurrence in nursing homes during the summer. A small child with diarrhea who has to rely on a parent to obtain water from a distant water source is also at risk. If a gastrointestinal infection also induces vomiting, which is typical of cholera, it makes it very difficult to drink enough fluids to replace the loss of water from diarrhea, even if an adequate source of oral fluids is available. Severe dehydration reduces blood volume, which decreases blood pressure. If untreated, the circulatory system collapses, resulting in clinical shock (very low blood pressure, reduced tissue oxygenation, lactic acid accumulation) and, ultimately, death.

The gastrointestinal tract is a veritable reservoir for body water. Normal daily digestive secretions from the stomach, pancreas, small intestines, and liver amount to between 8 and 10 liters a day in an adult of average size. More than 95 percent of intestinal fluid is reabsorbed. If the intestinal water-absorption mechanism is blocked, as occurs in infections by certain microbes, an average-size adult can easily lose several liters of fluid a day in diarrhea. Some infectious microbes produce toxins that increase intestinal secretions, in which case, water loss can double or triple. Because water in the intestines is in dynamic equilibrium with the plasma portion of blood, severe diarrhea can draw fluid out of the circulatory system into the gut, which can cause dehydration and circulatory collapse. This downward spiral is a physiological fact of life for the millions of children who die of diarrhea each year in the modern world, and the billions who have died prematurely throughout human history. The unimaginable loss of young children from watery diarrhea has been caused by many different viruses and bacteria; the most notorious of which is the bacterium that causes cholera.

CHAPTER **8**

The Cholera Morbus

Vibrio cholerae causes the most severe diarrhea of any microbe. An infected adult can easily lose 20 liters of electrolyte-rich intestinal fluid a day, and reportedly as much as 80 liters, rapidly bringing on severe dehydration. The patient's eyeballs shrink and appear buried in their sockets, the lips become parched, the extremities cold. There is a profound thirst that cannot be satisfied. Dehydration is so severe that 19th-century doctors observed a thickened, syrupy consistency of the blood. Without treatment or a spontaneous resolution of the problem, there is a rapid progression to circulatory collapse.

Cholera outbreaks came and went abruptly, like a force of nature, devastating one city, then moving on to the next. The mortality rate in the past was about 50 percent. The disease is particularly lethal in children under the age of five. To be in the midst of a cholera epidemic is terrifying. Adults who normally appear vigorous and healthy could turn into dried-up, emaciated corpses in half a day. A worker going about his business in the morning might return home later in the day to find his wife and children on the verge of death. It was horrible to witness, not only for the speed with which it takes lives but also for the gruesome nature of the victim's demise, leaving a cold shriveled body lying in soiled linens. The only reward for surviving the ordeal was protection against future *Vibrio cholerae* infections.

The onset of cholera as an infectious disease is not clear, but descriptions of a cholera-like illness appear in the writings of Nei Ching, who practiced medicine in China around 2700 BCE, and of

Hippocrates in Greece in the 5th century BCE. Sporadic outbreaks appeared throughout India and Europe in the 17th and 18th centuries; and the disease achieved worldwide prominence in 1817 when the first of several pandemics (global epidemics) began in India.

The opportunistic threads that tie together each pandemic are disease propagation from large gatherings of people housed under poor sanitary conditions, and its subsequent spread along trade and transportation routes. Military campaigns, refugee camps, fairs, markets, caravans, and religious pilgrimages have all inadvertently spread the illness. India's Ganges River, where millions of Hindu pilgrims assemble annually to cleanse themselves in its waters for the "ablution" of their sins, has been the site of many cholera outbreaks. Hindus hold their festivals in one of four cities every twelve years, depending on the alignment of the planet Jupiter and the Sun in the constellation Leo. In 1783, during the festival held in the city of Hardwar, twenty thousand pilgrims died over the course of eight days, an occurrence attributed to drastic temperature changes, going from hot days to very chilly nights with heavy dew. In the 1860s, cholera killed one-third of the 90,000 pilgrims who traveled to Mecca, the holy city of Islam, after which the disease spread to the rest of the Middle East. Between 1850 and 1931, there were 27 epidemics of cholera from such pilgrimages.

As many as 10,000 British soldiers stationed in India died during the first pandemic; no one knows how many Indians came down with cholera, but by extrapolation, the numbers must have been in the hundreds of thousands. The British then spread the cholera to Oman on the Arabian peninsula, where troops were sent to disrupt the slave trade. The disease also spread along trade routes to Iran and up the Volga River to Russia, as well as east from India to Nepal, Singapore, Java, and Borneo, and throughout Indonesia. From Java, the infection was carried into Japan, justifying the island nation's traditional desire for isolation.

In the 1830s, the second pandemic reached Russian soil, where it survived a harsh winter; the cholera-causing agent is able to survive bitter cold and tropical climates alike. Unknown to people living during that time, fecal matter from infectious diarrhea casually thrown onto the ice and snow could contaminate the residents of a community as the spring thaw reawakened the dormant killer, by bringing melted snow into the supply of drinking water. If only they had known what Dr. Robert Gilman, professor of epidemiology at Johns Hopkins, advised his students more than a century later: "The first rule of public health—remember, shit runs downhill."

Russian soldiers who were sent to stop a rebellion in Poland carried the disease west. Eventually 2.5 percent of the population of Moscow and St. Petersberg died of cholera, as did 3.5 percent of Germans and 5 percent of Hungarians. The epidemic emptied the streets of Europe's cities, people were so fearful of becoming infected. In Paris, where 96 of the first 98 cases died, the usually busy streets seemed nearly deserted, with few carriages visible, and the only stores doing much business were druggist's shops. Dead bodies had to be sewn in sacks because of a shortage of coffins. In some areas, cemeteries were overwhelmed with bodies, and the dead were buried on top of the dead.

Conditions were ripe in Europe for the spread of cholera. In cities, sanitation was poor or nonexistent. A lack of public facilities forced people to relieve themselves on stairs and streets; the latter were filled with a mixture of animal and human waste, washed only by heavy rainfalls. Nauseating gases emanated from the sewers. Drinking water had to be carried back to homes in pails, and the water was often discolored and foul-smelling.

Eventually, cholera crossed the English Channel and arrived in Britain. Although quarantine was considered to keep the disease off the British Isles, merchants and officials were against the idea, fearing that it would impede British trade. Dr. Thomas Latta, a British physician, administered intravenous fluids to fifteen moribund patients. However, only five survived, most likely because the patients were too critically ill when treatment was initiated. Although the experimental procedure was proclaimed to be life-saving by the journal *Lancet*, most establishment physicians were critical because it violated standard principles of treatment, which at the time included leeching, emetics, and cathartics. More than fifty thousand people in Great Britain died.

Irish immigrants eventually brought cholera to the United States. Many of them got sick en route because of overcrowding and unsanitary living conditions on the ships. Because the epidemic in Europe was expected to arrive eventually in the

United States, quarantine was attempted in the port cities where the ships arrived, but it was impossible to enforce consistently and was therefore ineffective. Immigrants would leap from boats before docking in order to avoid armed militia guards awaiting them. Thousands died of cholera soon after arriving, but not before spreading the infection throughout North America.

There were so many casualties in some cities, the dead had to be buried in mass graves. Of New York City's population of a quarter million people in the summer of 1832, 3,000 died—more than one percent. The number of deaths from cholera during the New York epidemic exceeded by three-fold all other causes of death. Walter Browne, the mayor of New York, in an act of desperation, proclaimed August 3, 1832 to be a day where the people of New York would unite to implore "the mercy of God to stay the pestilence that afflicts New York City." Newspapers advised that to avoid the cholera, one should lead a moral and temperate life. Those who could afford to leave fled the city. For the first time in decades, the streets of New York were quiet.

Another serious epidemic originated in New Orleans in 1842. From there it spread up the Mississippi to St. Louis where 4,000 people (about 10 percent of the population) died in eight months. Later, cholera was carried north and west by gold prospectors, steamboat passengers, and army troops. The Eighth U.S. Infantry stationed in the West lost one-third of its troops. Plains Indian tribes suffered heavy losses, especially those along the Oregon trail, where contact with whites was common. The Cheyenne tribe lost as many as 50 percent of their people. Only the Pacific Ocean stopped the westward spread of cholera.

During the American Civil War, cholera and other diarrhea-causing infections (euphemistically referred to by the soldiers as "the Tennessee trots," "the Tennessee quickstep," and "the green-corn rumbles," and by British troops in India as "Delhi belly") claimed 21,000 lives, about 3 percent of all casualties. Conditions on the battlefields were filthy. The water supplies were barely drinkable by humans or animals; soldiers had trouble getting their horses to swallow the foul water. By the second year of the war, according to an account written by G. W. Adams, the diarrhea/dysentery rate was 995/1,000. Confederate general Robert E. Lee lamented that his soldiers were worse than children at keeping clean.

Even the most remote regions on earth were not safe. In 1853, a whaling crew stationed in Davis' Strait and Baffin's Bay in the arctic circle recorded that an epidemic of the disease was devastating the local Eskimo population.

Although clearly an epidemic disease, the spread of cholera did not fit the pattern usually seen with contagious diseases such as measles and smallpox. Whereas most epidemics caused by human-to-human contact would gradually crescendo to a peak as each individual new case spread the disease to their contacts, cholera exploded onto the scene, rapidly causing many cases in a short period of time. However, families living together might fall victim while others who came in contact with them did not. Doctors treating victims rarely came down with the illness, in contrast to smallpox, where the risk of contagion was so great that voluntarily administering to the sick was considered heroic. And although doctors treating cholera did not usually fall to the illness, hospital washerwomen who handled soiled linen did.

Most medical doctors at the time clung to the idea that cholera was caused by a miasma, foul air emanating from decaying organic matter, or atmospheric emanations arising from the sick. In a boardinghouse in New York, several members of a family and a servant were stricken with cholera, killing two of them. Inspectors believed that a large quantity of cattle hides in an adjacent warehouse created a cholera miasma, having discovered that the lower portions of the hides were wet and in an "offensive state."

Rank, airless cellars, which served as homes for nearly 20,000 people in New York City during the 19th century, were viewed as brewing sites for the cholera miasma.[16]

---

[16]Cholera was only one of the problems for these poorest of the poor. Typhoid fever, smallpox, dysentery, and tuberculosis were rampant. The progressive decay of people living under these conditions was called "Tenant-House Rot." Premature aging developed in people as young as 30; the infant mortality rate was more than 30 percent, and the overall sickness rate was between 50 and 70 percent. Children with congenital deformities were ever-present. Old, unwashed bedding from dead smallpox victims was sold to rag-men for re-sale. Bedraggled street vendors with shedding smallpox scabs sold cigars and fruit to their impoverished neighbors.

In a boy's school near London in 1829, a foul drain was opened into a nearby garden and within two days, 20 of 30 boys were stricken with cholera. Doctors said that the "effluvia" emanating from putrid matter produced the disease.

Followers of the miasma theory believed that fresh air was the best way to prevent the disease. In Russia, it was observed that villages deep in the forest were free of cholera, attributed to the capacity of fir trees to neutralize the unknown cholera-causing agent. Yet, the disease often struck pristine settings. And, the miasma theory failed to explain why epidemics moved from east to west, against the trade winds. Also, in India, the disease did not follow the monsoon season, which, it was reasoned, should have increased the miasma potential. Some conjectured that cholera transmission was "telluric," coming out of the earth, or that emanations from the sick, influenced a terrestrial material that generated a "cholera poison."

Cultural and political differences fueled strange cholera theories. Paris, a city where one of every forty citizens died of cholera, was considered a "strong city," the air filled with deadly impurities. The British considered the French to exist in a low state of civilization with respect to bodily health, making them prone to cholera. The British described themselves as possessing a "greater tenacity of fiber, more vigorous flow of blood, and more copious supply of nervous fluid," which they believed made them more resistant to the disease. Poverty and debilitating lifestyle were thought to be contributors, along with sexual promiscuity, drunkenness, and gluttony. Religious practices were invoked to explain the high death rate in Paris and Rome, with Protestants in England attributing severe outbreaks of cholera to the "insalubrious food of the inhabitants of France and Italy" and their many religious fasts.

Among the conditions suggested as inducing cholera attacks were cold nights followed by hot days, or drinking cold water when the body is heated, or ingesting food or liquid that irritated stomach and bowels. Exposure to cold, exposure to sun, eating too much, having too little to eat—each theory had its proponents. After an attack of cholera among employees at a county jail in Fairhold, Connecticut, the illness was attributed to the "eating freely of cucumbers."

A weekly medical newsletter written during the 1832 epidemic in New York, the *Cholera Bulletin,* disseminated information about the disease, and, in what would be a modern breach of confidentiality, the addresses of infected households. But, nothing useful could be discerned about the spread of infection other than the observation that cholera "travels as man travels, stops where he stops, and proceeds again at the time and in the direction in which he resumes his journey."

People were desperate for answers.

Just as theories about the cause of cholera abounded, so did concepts regarding proper treatment. There were proponents and practitioners of bleeding, blistering, and leeching (leeches were applied to the anus for pain in the lower abdomen, or to the upper abdomen for pain in the stomach); of poultices administered to the abdomen; of ice administered internally. Some prescribed opium to quell the diarrhea. One noted physician in 1669, Dr. Thomas Sydenham, encouraged the use of cathartics to treat cholera to "expel the sharp humours that are the fuel of the disease."

There were those who prescribed tobacco and brandy, and those who recommended decreased or increased nourishment. Others recommended that the air be purified by lighting bonfires, burning tar, and firing off great guns. During the 1865 cholera outbreak in Paris, while Pasteur and his followers were elucidating the germ theory of disease, less knowing citizens exploded sulfur in the air to clear it of the "cholera poison." Quacks abounded. Various nostrums, pills, salts, potions, mustards, and plasters were announced as potential cures. One entrepreneur, who called himself the Fire King, swallowed an "elixir cholerae" of phosphorus, prussic acid (hydrogen cyanide), and live coals to combat cholera, which he sold at the exorbitant price of ten dollars. To promote his product, the Fire King used the names of prominent New York physicians in his advertisements, without their permission or approval.

The cause of cholera was revealed in the course of three decades of scientific observations beginning in the 1850s. During the epidemic that struck Florence in 1854, an Italian physician, Filippo Pacini (1812-1883), dissected the intestines of recently deceased victims, and observed in them the curved bacterium that causes cholera, after microscopic examination; he named the comma-shaped organism *Vibrio*. His report was completely ignored, since the germ theory of disease had not yet been established, and his findings were lost to the scientific world for decades. After detecting the *Vibrio* bacterium, Pacini continued to do important research in cholera and even recognized the key role played by salt and fluid loss in the disease's progression. He died poor and largely unrecognized, having spent his own money for research.

The work of John Snow in 1854 revealed the true nature of cholera's contagion. Snow (1813-1858) was a prominent physician who had risen from a working-class background to become one of the first anesthetists in Britain. He was an early experimenter with chloroform, often using himself as a subject for this risky general anesthetic. Snow helped establish the usefulness of chloroform in obstetrics when he successfully administered the anesthetic to 77 women in labor during the late 1840s. Chloroform-assisted deliveries gained wider acceptance, especially among the social elite, after Snow was granted permission to chloroform England's Queen Victoria for the last two of her nine deliveries.

Snow first became interested in cholera in 1831 while working as an apprentice to a doctor. He observed that miners working underground came down with cholera–their exposure in an area free of swamps and organic decaying matter contradicted the prevailing miasma theory. Later, during a cholera epidemic that struck England in 1848, Snow made important observations about cholera's transmission. It was during this epidemic that a well-publicized outbreak occurred at an asylum for children in the village of Touting, near London. In 1834, the British government, in a desperate legislative effort to deal with the ever-increasing poverty among the "lower classes," passed the notorious amendments to its "poor laws," making a distinction between the working poor and the non-working poor. One of the consequences was the separation of children from their parents, under the assumption that non-working poor adults were unable to care for themselves, and therefore were incapable of properly caring for their children. The parents were sent to poorhouses, the children to pauper asylums. At the Touting asylum, 1,400 children were placed under the care of Bartholomew Peter Druet, who was given a stipend for each child, an incentive that encouraged overcrowding. Conditions there were deplorable, with the children dressed in rags, even in cold weather, and receiving too little food. During a two-week period in January 1849, about 200 children at the asylum died of cholera, although a nearby insane asylum and the town of Touting itself were unaffected. Criminal charges were brought against Druet when the awful living conditions were uncovered during an investigation of the cholera outbreak. Newspapers closely followed the court case, providing vivid daily accounts of the proceedings. One of the reporters was Charles Dickens, who accused Druet of "shameful indifference and neglect." The episode set the stage for changes in the poor laws and more progressive social reforms. In later years, an English minister would claim that the great social agencies in England were "the London City Mission; the novels of Mr. Dickens; and the cholera."

While studying the outbreak, Snow came to the conclusion that cholera began with the ingestion of a contaminant that directly affected the intestines. From an analysis of individual cases, he concluded that cholera was caused by an agent that grew in the body and was transmitted to others by vomit or stool that contaminated drinking water. He published his observations in 1849 in a pamphlet called "On the mode of communication of cholera." London was in the second year of a serious epidemic during which tens of thousands were killed. But, his conclusions were ignored.

In 1854, Snow got the opportunity to put his theories to a test when London was struck by another cholera epidemic. He showed that in one district of south London, the cholera death rate was nearly 15 times greater in homes supplied by the Southwark and Vauxhall Water Company compared with its competitor, the Lambeth Water Company. Both drew water from the Thames, but the Southwark and Vauxhall Water Company site was known to be contaminated with sewage; the Lambeth Water Company had recently moved to an area above the contamination. Officials dismissed the observation believing that the great volume of the Thames River would dilute out any cholera-causing poison.

One very serious outbreak occurred in a quarter-mile region in London's Soho district. Within 10 days, 500 people in the area died. Snow suspected that the concentration of deaths in such a localized region was due to contaminated water. The Soho district was supplied by a much-used pump at the corner of Broad Street and Cambridge which was attached to a well below, which, in turn, drew water from the Thames. There were reports early in the epidemic that water from the pump had developed a mildly offensive odor. He investigated all cases using the City of London Registry of Deaths and found that the majority of cholera victims lived within a few hundred yards of the pump.

There were other circumstances that led Snow to suspect the Broad Street pump. Nine customers of a coffee shop located within yards of the pump died in the early days of the outbreak. In a nearby workhouse, there were only five deaths among 535 inmates, and yet residents in surrounding houses were devastated by cholera. Snow discovered that the workhouse had its own well. In a local brewery, there were no deaths among the workers, whereas a neighboring factory reported numerous deaths. Snow found that the brewery workers drank beer instead of water, while the factory workers drank from a common trough filled with water from the Broad Street pump. He also investigated several deaths that occurred miles from London and found that many victims had water from the Broad Street site shipped to their homes because the taste was preferable to water from other pumps.

Snow presented his findings to the St. James Parish Board of Guardians and suggested that the pump handle be removed. The board consisted of prominent members of the parish and served as a de facto local government. Snow's views were met with disbelief: not a single member agreed with him, including the one doctor in the group. Nevertheless, the board had the pump handle removed, acquiescing out of panic and a sense of frustration rather than being persuaded by Snow's observations. The Soho outbreak quickly subsided.

Snow suspected that sewage had seeped into the well. He discovered that a few days before the Soho outbreak, an infant girl had died of cholera–her family lived in a house adjacent to the Broad Street pump. Snow's colleague, the Reverend Henry Whitehead, interviewed the baby's mother, Sarah Lewis, and discovered that she had cleaned dirty diapers in a pail and emptied the residue into a cesspool in front of her house, a few feet from the pump well. The cesspool was excavated; poor construction had resulted in a backup of sewage which leaked through cracks in the brickwork into the pump well.

Snow's persistent detective work, which launched the field of epidemiology, demonstrated beyond doubt that cholera was transmitted by human waste. However, the General Board of Health dismissed the findings. It was only after the germ theory of disease was established in the coming decades that Snow's observations were finally accepted. Unfortunately, he did not live long enough to fully appreciate what he had accomplished. Snow died of a stroke in 1858, at the age of 45, a premature death brought on, according to some of his friends and colleagues, by his experiments with chloroform, a strict vegetarian diet, and abstinence from alcohol.

Robert Koch (1843-1910), the founding father of modern microbiology, along with Pasteur, rediscovered *Vibrio cholerae* nearly three decades after Pacini first detected the microbe. Koch became interested in bacteria as a cause of contagious disease after encountering the early work of Pasteur (who was two decades older). He began his investigations quite modestly in the Wollstein district in Germany, in a home-built laboratory using a microscope given to him as a present from his wife. He found time from his private practice to conduct laboratory research and refine his microscope skills. As his interest in bacteria grew, Koch began to refer his patients to other physicians and eventually settled into a full-time research career. Koch's initial breakthrough finding was on anthrax.

During the 1870s, farmers in the region were losing animals to anthrax and suffering great economic loss. Some farmers were coming down with the illness as well. In 1850, Aloys-Antoine Pollander concurrently with Pierre-Francoise Olive Rayer and his disciple Casimir-Joseph Davaine had discovered a bacillus bacterium from an anthrax-infected animal, but its role in the infection was not conclusively established. The presence of bacteria does not necessarily imply causality; cultures from the throat, sputum, skin, and feces will grow many types of bacteria, most of which are harmless. Koch, convinced that the bacillus they had found did cause anthrax, set up a series of simple yet meticulously designed experiments to prove it. He successfully infected mice with anthrax by inoculating them with material from the spleens of infected animals; spleens taken from normal animals did not cause the disease. Koch found an effective way to culture and purify the anthrax bacillus. He was then able to cause anthrax in mice using bacterial strains that had been grown for several generations in the laboratory, completely bypassing the natural animal reservoir. Although the idea of causing infection with a laboratory strain of bacteria is taken for granted today, the experiment was an eye-opener in the nineteenth century. Growing microbes in the laboratory and causing a predictable disease by injecting it in animals was a revolutionary experimental finding in Koch's time. His presentation to establishment scientists and his published report in 1876 were so convincing that the discovery caused a sensation, despite Koch's lack of credentials and formal research training. The bacterial cause of disease, an idea initially proposed by Pasteur, was now on firm scientific footing. The field of bacteriology was born. Within two decades, the bacteria responsible for the most important infectious killers of mankind were identified, primarily by Koch and his associates.

The anthrax findings also provided the foundation for the scientific standards needed to establish associations between a disease and its putative microbial cause, a set of principles known as Koch's Postulates that still guides the field of infectious disease medicine. Koch suggested that in order to prove causality, a specific bacterium must be present in an infected person (first postulate); the bacterium must be isolated from the diseased host and grown in pure culture (second postulate); and the specific disease must be reproduced when a pure culture of the bacterium is inoculated into a healthy susceptible host animal (third postulate). However, these rigid guidelines do not apply to every infectious organism; microbes that cause disease in humans are not always infectious in laboratory animals.

Over the next few years, Koch revolutionized medicine with his discoveries of the causative agents of tuberculosis (the single greatest killer of adults during the 19th century) wound sepsis, and many other infectious diseases. He and his staff also invented experimental techniques still used in clinical and research microbiology, including a method for staining bacteria (the Gram stain, named for the assistant who developed the technique) and a process for culturing individual colonies of bacteria on agar (a firm, easy-to-prepare gel). Koch spent years trying to find a good method for growing bacteria on a convenient surface. For some time, his laboratory used gelatin containing beef broth. However, gelatin melts at high ambient temperatures, and some bacterial strains can digest it as a food source. It was Fanny Eishemius, the wife of a laboratory assistant, who offered a solution. She had learned about the use of an extract of seaweed called agar, a complex polysaccharide that dissolves in boiling water and forms a solid below 45 degrees centigrade. Fanny used agar to make jellies. Solidified agar mixed with nutrients turned out to be the ideal matrix for growing bacterial colonies. With a few modifications, this is essentially the same technique used today to isolate bacterial strains.

Koch combined his extraordinary laboratory skills and love of travel to discover the cause of cholera. He led a team of German physicians and scientists on an expedition to Egypt and India during a severe outbreak there in 1883. Troops returning from India on British ships probably introduced the cholera to Egypt after passing through the recently completed Suez Canal. As many as 50,000 Egyptians died of the disease in the summer of 1883. The Germans had competition from a French team dispatched to the region by Pasteur. However, the French gave up their research after a member of their team contracted the illness and died. Koch was able, after some difficulty, to isolate a comma-shaped organism, *Vibrio cholerae* (Kommabazillen in German), from infected individuals and grow it in pure culture.

However, Koch was unable to infect other animals with cultured *Vibrio cholerae*, thus failing to satisfy his own third postulate. He reasoned correctly that his laboratory animals were resistant to the bacterium. Despite this problem, Koch was able to convince most skeptics by showing that the bacterium could be isolated from all cholera victims, as well as from local drinking water. However, his difficulties in satisfying his own third postulate for *Vibrio cholerae*–a consequence of the microbe's narrow host range–provided critics with ammunition against the germ theory of cholera. A prominent Bavarian scientist, Max von Pettenkofer, agreed that a cholera germ originated in India, but he believed that it had to come in contact with a local factor present in the soil in order for the cholera toxin to cause disease. He published his strange theory in 1869, in the very first issue of the journal *Nature* (now the most prestigious scientific journal in the world). He strongly objected to the idea that cholera could be transmitted by drinking water and vehemently rejected Koch's findings. To prove the point, von Pettenkofer reportedly drank excreta from a cholera victim and somehow avoided contracting the illness (or he lied about drinking the nasty concoction).

Prominent British physicians, commissioned by the secretary of state for India, Randolph Churchill (father of future prime minister Winston Churchill), prepared an official refutation of Koch's findings. There was a strong economic incentive to reject the idea of a cholera germ carried from India to other parts of the world, considering the benefit to British commercial interests garnered from the completion of the Suez Canal, which decreased the time and expense of transporting goods between India and Europe. Surgeon General W. G. Hunter, a British official dispatched to Egypt to study the epidemic, claimed that unusual weather conditions had brought cholera to Egypt, a most satisfying theory for the British government.

Eventually, *Vibrio cholerae* and its transmission through contaminated water supplies were accepted as the cause of cholera. The findings helped provide the impetus to bring clean water into cities and to develop an infrastructure for waste disposal and sanitation. These developments and the inventions of the flush toilet and indoor plumbing resulted in the virtual elimination of cholera from developed countries. Koch was awarded a Nobel Prize in 1905, more for his complete body of work in bacteriology than for cholera specifically.

The bacterium that causes cholera is a member of a large genus of organisms commonly found in fresh- and salt-water estuaries and coastal waters throughout the world. They are free-swimming organisms that use a single long flagellum for motility, similar to sperm. *Vibrio cholerae* also exists in nature as biofilm aggregates, adhering to various sea creatures. The bacteria bind to chitin, a mucopolysaccharide (a type of carbohydrate) found at high concentrations in the shells of crustaceans. The hindgut of crabs, planktonic crustaceans, the egg sacs of copepods, and bivalves (oysters, clams, and mussels) are some of the organism's habitats. *Vibrio cholerae's* use of oysters as a refuge is a property shared by its cousin, *Vibrio vulnificus*, the microbe responsible for one of the deadliest forms of food poisoning, with a mortality of approximately 50 percent (*Vibrio vulnificus* is responsible for more than 90 percent of fatal food poisoning caused by sea food). *Vibrio cholerae* can also bind to inert suspended matter in water, as well as to the roots of water hyacinths. Eating raw bivalves and undercooked shellfish are vehicles for transmitting *Vibrio cholerae*, but these routes are much less common than drinking contaminated water.

Vibrio cholerae is killed by heating water to 60 degrees centigrade. However, heat sterilization is difficult in some parts of the world where firewood and oil are scarce and expensive. Consequently, researchers and public-health officials are always searching for simple ways to combat the bacterium that can be implemented in poor nations with limited resources. Water stored in a discarded plastic soda bottle, painted black on one side to reflect infrared radiation from the sun, like a portable outdoor shower bag, can become sufficiently hot to kill the offensive bacterium. Adding chlorine to water is effective, and so is lime juice, since *Vibrio cholerae* is very acid sensitive.

Filtering *Vibrio cholerae* is also an effective preventive strategy. Clever researchers recently found that water can be filtered free of the bacterium after passing through the material used to make a sari.

The strongly acidic gastric secretions found in the human stomach offer a high degree of protection against *Vibrio cholerae*. However, the ability of *Vibrio cholerae* to form a biofilm aggregate provides a measure of protection to the bacteria as they pass through the acid-filled stomach. Acid production is low in elderly people, infants, and in the malnourished, resulting in an increased cholera risk. Tens of millions of people are taking medications for peptic ulcers and heartburn that suppress stomach acid output, putting them at theoretical risk for cholera, should they ever ingest *Vibrio cholerae* while traveling to an endemic area.

After they are ingested, the surviving *Vibrio cholerae* adhere to cells lining the wall of the small intestine. Once the organism proliferates, it produces the toxin that causes intestinal loss of water and electrolytes. Approximately two dozen *Vibrio cholerae* genes are engaged in the coordinated production of the various proteins involved in intestinal adhesion, colonization, growth, and toxin production, the "pathogenicity factors" responsible for infection. Most of the pathogenicity genes encoding these factors were acquired through genetic transfer of DNA from unknown microbes via two bacteriophages, events that probably occurred within the past few thousand years. The most important pathogenicity gene is the one that codes for the toxin protein itself (*ctxA*).

Cholera toxin induces its deadly effect by disrupting signal transduction, the process that cells use to communicate with the external world and deliver information throughout the cell. Signal transduction is a multistep process that converts a biochemical signal, such as the influx of calcium or the binding of a growth factor or hormone to a membrane receptor, into useful work by the cell. Cells are very active in their baseline state, but they await instructions from stimulated signal transduction pathways in order to do something grand, such as undergo cell division, step up antibody production, or build a stronger synapse in the brain. Almost every programmed cellular function–the opening of an ion channel, DNA replication, gene activation, the ability of phagocytic cells to engulf bacteria, activation of nerve and muscle cells, the fertilization of an egg by sperm, to name but a few–involves one or more of the different signal transduction pathways that exist in eukaryotic cells.

An essential feature of signal transduction pathways is their multistep activation scheme. Step one activates step two which activates step three, and so on, all the way down the line until the signal reaches its ultimate target, such as the cell nucleus or an ion channel. Another feature is that at each new activation step, the previous step is usually inactivated. Think of a series of switches with each one turning on the switch in front, while at the same time turning the switch behind it to the off position. The multistep activation scheme and the feedback negative controls have been inserted during evolution to provide cells with a mechanism for exerting tight control over signal transduction. Without controls in place, cells would run amok. Cell growth would continue unchecked, causing multiple cancerous growths, and the brain would get so big laying down countless useless synapses, it would smash itself against the skull.

Because of the importance of signal transduction in maintaining normal cellular function, it is no wonder that many diseases, from major depression to cancer, occur when the process is disrupted in some manner. It is also not surprising that many microbes have evolved with genetic mechanisms in place to manipulate host signal transduction pathways for their own destructive ends. *Vibrio cholerae* is one of the cleverer microbes in this regard. Its poisonous effect on the human gastrointestinal tract is caused by disrupting one of the most important signal transducing molecules, the G-protein. G-proteins lie dormant on the inner portion of the cell membrane, adjacent to receptor partners. Hundreds of different "G-protein coupled receptors" (GPCRs) are encoded in the human genome, indeed, the genomes of most complex organisms, all genetic cousins of one another, related to a primordial GPCR that appeared at the dawn of eukaryotic evolution. They carry out diverse functions in cells; from regulating smell and vision, to controlling neuronal transmission and endocrine gland function. When G-proteins are activated, following stimulation of their adjacent GPCR, they perform some useful work for the cell, one of which is to increase the level of the intracellular regulatory molecule, cyclic AMP (cAMP).

G-protein activation is brief, lasting only seconds or minutes. An intrinsic, self-regulating feature built into each G-protein keeps it from running amok in the cell. This brake, an enzyme called GTPase, is blocked by cholera toxin. Several G-proteins are potential targets of cholera toxin, but the one that activates cAMP in the gut is exquisitely sensitive. By inactivating GTPase, the G-protein is placed in a state of constant activation, like a light switch stuck in the on position, leading to a dangerous accumulation of cAMP. Cyclic AMP regulates fluid and salt flow from the cells lining the gut. When cAMP accumulates above its normal physiological level, water and electrolytes flow out of the cell uncontrollably. The loss from any individual cell is minuscule; but, multiplied by the hundreds of billions of cells lining the intestinal lumen, cholera toxin poisoning leads to a huge loss of water and electrolytes that leave the intestines in a rushing stream. The evolutionary strategy used by *Vibrio cholerae,* when it acquired a bacteriophage containing the *ctxA* gene long ago, was the ability to spread its progeny by the toxic disruption of G-protein's regulation of water and electrolyte levels in the human small intestine, inducing copious watery diarrhea loaded with infectious bacteria.

Much of what we know about G-proteins today is due to our exploitation of cholera toxin as a research tool. Since G-proteins are activated for a very brief period of time, it is difficult for researchers to study their effect on signal transduction pathways. By treating cultured cells with cholera toxin, a prolonged effect on signal transduction can be induced under laboratory conditions, enabling scientists to study the process more readily. The analysis of signal transduction using cholera toxin has aided the discovery of drugs used in treating depression, heart disease, asthma, and cancer.

There is evidence that the mutations responsible for cystic fibrosis expanded in the population because they provide protection against cholera and other causes of secretory diarrhea. As discussed in chapter 10, cystic fibrosis is a severe genetic disorder caused by inheriting two mutated versions of a gene called *CFTR*, one from each parent. It is the most common fatal genetic disorder in people with Northern European ancestry. *CFTR* codes for the cystic fibrosis transmembrane regulator, a cAMP-activated ion channel that transports chloride and bicarbonate across cell membranes. There is evidence that carriers of cystic fibrosis, who have a single copy of mutant *CFTR*, experience reduced salt and water loss when exposed to cholera toxin and other bacterial toxins that increase cAMP levels in the gut, thereby reducing diarrhea and improving the chances of survival. In a mouse model of cystic fibrosis, the rodent carriers of the disease were found to excrete 50 percent less fluid in response to cholera toxin. Intestinal biopsy samples from humans with *CFTR* mutations have also shown a reduction in fluid secretion in response to various cAMP activators.

One problem with this hypothesis is that cholera did not reach Northern Europe–the population with the highest frequency of *CFTR* mutations–until the early 1800s. Given a *Vibrio cholerae* infection rate of about 10 percent, the period of time that cholera caused serious disease in Europeans was too short for *CFTR* mutations to expand by Darwinian selection. One possible explanation for this paradox is that *CFTR* mutations initially expanded in Caucasians because of resistance to other common causes of watery diarrhea. The toxin produced by diarrhea-causing strains of *E. coli* produces loss of electrolytes and fluids by activating the *CFTR* channel, albeit by a different mechanism than cholera toxin. Although the prevalence of *E. coli*-induced diarrhea in the past is not known, it is far more common in the modern world than cholera. Assuming a similarly high prevalence in the ancient world, it is possible that *CFTR* mutations expanded in the population over thousands of years by affording some protection against *E. coli*-induced diarrhea. Another candidate for the expansion of *CFTR* mutations in Europeans is typhoid fever, which will be discussed in the next chapter.

If mutations in *CFTR* initially expanded because of other infectious diseases, they may have been fairly common in European populations by the time cholera arrived. A small percentage of Northern Europeans may have escaped a cholera death by riding on the genetic coattails of other infectious diseases.

Recently, investigators identified a drug inhibitor of the *CFTR* channel called thiazolidinone. This drug effectively blocks cholera toxin's induction of intestinal fluid secretion and is being tested as a treatment for watery diarrhea caused by *Vibrio cholerae* and *E. coli*. Thiazolidinone mimics the effects of having a defective *CFTR* gene, and its ability to block the action of cholera toxin support the notion that mutant genes that can cause cystic fibrosis protect against cholera and other secretory diarrheal diseases.

Before the research by Snow and Koch, no significant survival pressure was exerted on *Vibrio cholerae*. With humans living under squalid conditions, in utter ignorance about the fecal-oral spread of disease, the cholera bacillus was left undisturbed. It was able to create as many as a quintillion copies of itself throughout the world. Since there was no significant selective pressure exerted during the first six cholera pandemics, there was no compelling pressure for *Vibrio cholerae* to change.

However, by the end of the 19th century, the findings by Snow and Koch began to have an effect. The need for clean water and effective sewage disposal was recognized by municipalities. The increasingly widespread use of flush toilets, sewage systems, water filtration, heat sterilization, and improved (but not perfect) personal hygiene, began to put a damper on the bacterium's 200-year-old parade. Cholera nearly disappeared from the Western Hemisphere. The rapidly changing socioeconomic landscape and increased scientific knowledge were exerting selective pressure on *Vibrio cholerae* to change or face possible extinction.

The result was the emergence of a new variety of the bacterium. The first six pandemics were caused by a strain called 01. The new strain, El Tor 01, first detected in Indonesia in 1961, has developed an apparent counter-evolutionary survival strategy compared to its 01 counterpart: it's a weakling. Toxin production in El Tor 01 is expressed under more restrictive circumstances than the classical 01 strain, and with reduced efficiency. Infected people have less severe diarrhea and are therefore less debilitated. Although this change in the bacterium reduces the number of infectious particles expelled into human sewage, in the form of huge volumes of *Vibrio*-filled watery diarrhea, diminishing its ability to infect others through its most common route (via large-scale water contamination), the El Tor strain enhances its own survival by subtler means. By causing less debilitating diarrhea, it increases its capacity for direct person-to-person transmission. With only a moderate case of cholera, mothers and other people who handle food are less debilitated and, instead of being confined to bed, are more likely to be up and about while infected, allowing them to transmit the bacterium while preparing meals. Thus, El Tor 01 has managed to bypass improved sewage-management conditions in the world by exploiting lax personal hygiene and person-to-person transfer, rather than wide-scale transmission.

Additional selective pressure has been exerted on *Vibrio cholerae* due to the improved survival of cholera victims in the modern era, a direct consequence of Oral Rehydration Therapy and intravenous fluids. As the number of cholera survivors increases, there are more and more people alive who have developed immune resistance to *Vibrio cholerae* 01 strains (immunity to classical 01 also provides immunity to El Tor 01). As a result, a new strain of *Vibrio cholerae* emerged in 1991, called 0139. Its major evolutionary feature is the lack of cross immunity with 01 and El Tor, preventing antibodies to the classic strains from working against 0139. A mutation in a gene responsible for the surface polysaccharide expressed on the bacillus's outer membrane (a major stimulant of the human immune system) is responsible for this change.

Strains of *Vibrio cholerae* resistant to antibiotics are also beginning to appear. One strain found in nearly all isolates of *Vibrio cholerae* from Asia has acquired a transposon called *SXT* that contains resistance genes against all the major antibiotics used to treat cholera. *SXT* is also found in several other bacterial species, demonstrating the promiscuous nature of transposon-mediated DNA exchange. Curiously, *SXT* transfer between bacteria is augmented by the antibiotic Cipro, one of the world's most commonly used antibiotics. An unanticipated biochemical effect of the antibiotic is that it increases the expression of genes needed for *SXT* exchange. Thus, in some cases, the acquisition of resistance genes is not simply a passive process due to chance and Darwinian selection. The very antibiotic used to treat infectious diarrhea can itself shift the odds in favor of antibiotic resistance by directly increasing the frequency of DNA transfer.

Despite the ability of the crafty microbe to reconfigure, in the battle of the genomes between *Homo sapiens* and *Vibrio cholerae*, humans now have the upper hand. Most developed nations have been free of cholera for decades, and mortality has been substantially reduced. Although we can expect future pandemics–since the opportunistic *Vibrio cholerae* needs only a single infected individual living under conditions that support fecal contamination of drinking water to cause a large outbreak–the bacterium has been forced back, in developed nations at least, to its tranquil watery origins, hitching a ride on crab shells and clams.

CHAPTER 9

The Burning Fever

Along with the desire for a new life free from religious persecution, poverty, and famine, the earliest settlers to North America also brought one of Europe"s biggest health problems, typhoid fever. It was known as one of the three "burning fevers," together with malaria and yellow fever, a virulent trio of infectious diseases that nearly wiped out the Jamestown and Plymouth colonies in the early 1600s. During the nineteenth century, it became most commonly associated with the squalor of urban life–especially overcrowding and poor sanitation–and with the dirty living conditions of soldiers in the field. Typhoid fever killed tens of thousands of people every year in Europe and the United States until the turn of the 20th century. It struck 20 percent of the troops fighting in the Spanish-American War in 1898, justifying its dubious distinction as the "soldier"s disease." At the turn of the 20th century, a committee of the best infectious disease experts and epidemiologists convened to investigate the high infection rate in the military concluded that the four "f"s were responsible: feces, filth, flies, and fingers. Like cholera, typhoid fever is spread by fecal-oral contamination.

Until the 20th century, it was believed that flowing water was an effective method for cleaning human waste, which led to the habit of dumping raw sewage into rapidly moving bodies of water, with obvious consequences to people living downstream. Only after the means of contagion for both typhoid fever and cholera was established did social and political reform change the way municipal water supplies and sewage were handled. During the late 19th century, a number of programs were implemented through the conviction that public health measures initiated by progressive government could control disease. Sanitation departments were assembled in large cities to clean streets and remove trash; aqueduct systems were built to bring fresh water from rural regions into major cities; drinking water was filtered and disinfected with chlorine; health departments were created to investigate and contain outbreaks of disease; and bonds were issued to pay for them all. Change was slow and expensive, and not universally practiced. In 1900, the state of Missouri filed a lawsuit against Illinois to prevent the city of Chicago from dumping raw sewage into the Illinois River, which was causing persistent typhoid fever in St. Louis.

Although slow to come, there were measurable improvements. In 1890, only 1.5 percent of urban dwellers drank filtered or treated water; but by 1914, the number had risen to 40 percent. It took nearly half a century of continuous reform before these actions led to a substantial decline in the number of cases of typhoid fever in the United States, where eventually, the disease was driven to near obscurity. Today there are only a few thousand cases reported every year, mainly in foreign travelers returning home, or immigrants from endemic areas. Before the antibiotic era, the case fatality rate (the percentage of people who die after an illness) was about 15 percent. Nine years of piloting the first airplanes proved less risky to Wilbur Wright than ingesting the contaminated morsel that resulted in his deadly typhoid fever infection in 1912. Today, typhoid fever is primarily confined to underdeveloped nations, with nearly two million cases in the world every year and six hundred thousand deaths, a case fatality rate that is actually higher than the rate had been in the 19th-century Europe. Poor access to antibiotics in remote parts of the world and lack of adequate nutrition are the 21st-century culprits responsible for this persistently high mortality rate. In some villages in Papua New Guinea and Indonesia, 30 to 50 percent of infected people die.

Unlike cholera, severe diarrhea is not a feature of typhoid fever. Instead of causing dehydration and circulatory collapse, typhoid fever kills when the causative organism damages organs of the body, primarily the small intestine, and causes dangerously high, persistent fevers and delirium. After you contract cholera, it usually takes a day or two for your fate to become apparent. However, the effects of typhoid fever are more delayed and unpredictable; infected patients can linger for a month with the illness, sometimes appearing to improve, then suddenly die from a bowel perforation, one of the major complications of typhoid.

Typhoid fever is caused by the bacterium *Salmonella typhi*, which is one of many *Salmonella* species that can cause disease in humans. It is a distant relative of *E. coli.* DNA sequence analysis of the two microbes indicates they diverged from a common ancestor about 100 million years ago. *Salmonella typhi* is a selective human pathogen, unlike other *Salmonella* species, which can infect many mammals and birds as well as humans. It was discovered in 1880 by a Swiss bacteriologist, Carl Joseph Eberth, and confirmed by Robert Koch a year later. Although typhoid fever is rare in the United States today, other forms of *Salmonella* disease are not. *Salmonella* contamination is one of the most common causes of food poisoning in the world, with millions of cases in the United States alone, mainly from chicken and turkey products. Some common sources of outbreaks are picnics, street fairs, school cafeterias, communal pot-luck dinners, and street vendors, as well as the contamination of utensils and cutting boards by raw poultry and eggs, creating opportunities for the bacteria to end up in people's mouths. The overuse of antibiotics to prevent *Salmonella* infection on poultry farms has created resistant bacterial strains that have managed to find their way into human hosts, as mentioned in chapter 1.

Salmonella typhi is destroyed by chemical disinfection, heat, and light. Thorough cooking of contaminated foods can prevent infection. However, the bacterium survives freezing temperatures and can remain viable for weeks in contaminated cisterns, sewers, and soil. A variety of different antibiotics are effective against it, but drug resistance is emerging through its acquisition of plasmids that contain genes coding for resistance. One strain of *Salmonella typhi* has acquired pHCM1, a super plasmid with genes that confer resistance to four antibiotics commonly used to treat typhoid fever. Resistant strains are now found wherever typhoid fever is endemic–that is, everywhere in the world except the United States, Western Europe, and Australia.

Salmonella typhi is remarkable for its ability to thrive in some of the more bacteria-hostile places in the human body. First, it must survive passage through the acid-rich stomach. Then it infects phagocytic cells in Peyer's patches, the extensive system of immune cells located throughout the intestinal tract that plays such a critical role in destroying ingested pathogens. From there, it spreads throughout the body to infect other phagocytic cells in the spleen and blood, the very cells in the body normally relied upon to squash invading microbes. The bacterium can also remain in a quiescent state for years inside the gall bladder, an apricot-sized sack which is attached to the liver and which is filled with bile, the green chemical brew released into the small intestine to help it digest fatty meals. Bile is inhospitable to most bacteria. A comparison of DNA from *Salmonella typhi* and non-pathogenic strains reveals that the former has acquired several large pathogenicity islands through genetic transfer from bacteriophages and plasmids, supplied by unknown bacterial sources, which have provided it with the genes needed to adapt to the myriad of uninviting roadblocks put up by human hosts.

Like *Vibrio cholerae*, *Salmonella typhi* is acid-sensitive. When the bacteria are ingested in a bolus of food, the latter provides a protective barrier that enables some bacteria to pass through the stomach unscathed, allowing it to reach its primary target, the small intestine. Although millions of *Salmonella typhi* may be ingested in contaminated food or water, most are killed in the stomach. However, only a thousand organisms need survive transit through the stomach's acid in order to establish an infectious niche further down the digestive tract. There, they invade cells located on the surface and within the wall of the small intestine, one of which is bacteria-eating phagocytes that permeate throughout the inner wall of the gut. These cells usually function as sentinels waiting to engulf pathogenic organisms. But *Salmonella typhi*, haughtily, invade them instead. As the bacteria grow in the small intestine, they can erode the intestinal wall from the inside, causing a dangerous hemorrhage, or through the outside, perforating the intestine, spilling *Salmonella typhi,* and commensal bacteria that normally inhabit the gastrointestinal tract, into the abdominal cavity and causing a dangerous infection called peritonitis. Intestinal perforation from typhoid fever, a ruptured appendix, or any other source was always fatal before antibiotics were discovered; even Harry Houdini's magic couldn't save him from the effects of peritonitis caused by a perforated appendix in the pre-antibiotic era.

After colonizing the intestinal wall, *Salmonella typhi* is transported to phagocytic cells in the blood, bone marrow, liver, and spleen. The bacteria spread by killing and rupturing those infected phagocytic cells, spilling hundreds of millions of infectious organisms into the blood. *Salmonella typhi* destroys phagocytes by secreting proteins that induce them to undergo apoptosis, which is essentially a biochemical program that initiates a chain of events culminating in the cells committing suicide. Most cells in the body have a limited life span: they grow old, become senescent, and die. Dead cells are replaced by new, fresh ones, generated from various pools of regenerative stem cells. Under some circumstances, cells take a more active role in their own demise by initiating cellular suicide. Apoptosis is an important physiological feature in organ development and in keeping the immune system in check. During brain development, for example, many neurons shrivel up and die from apoptosis because the connections they create do not work, making them irrelevant. Without apoptosis, the brain would be grow well beyond its normal size, swollen with useless neurons, enlarging the head so much that newborns would get stuck going through the birth canal, and making us all look like Frankenstein's monster. The immune system uses apoptosis to destroy B- and T-lymphocytes that recognize and target an individual's own proteins. Immune cells that maintain their self-recognition capacity by escaping apoptosis can reemerge years later and cause autoimmune diseases.

The apoptosis pathway has also been established as a system for destroying unwanted cells that are damaged beyond repair, especially those that have sustained severe DNA damage, making them prone to develop into cancers. One of the pathways involved in the generation of cancer is the acquisition of mutations in apoptosis regulatory genes that limits a cell's capacity to destroy itself when severe DNA damage has occurred (see chapter 15). Apoptosis is nature's equivalent of pruning shears.

Apoptosis is initiated through pathways that activate proteins called "executioner caspases," which are enzymes that cleave other proteins. Caspases target and destroy vital proteins, which are needed to maintain cellular integrity by helping to stitch DNA together. The caspases turn the continuous three-billion-nucleotide string of DNA into useless fragments, preventing any possibility of escaping the death signal. Phagocytic cells mop up the damaged cells by recognizing a variety of surface protein alterations referred to as "eat me" signals, which are induced in dying cells by the apoptosis pathway.

Phagocytic cells themselves are subject to apoptosis when they are damaged by the various bacteria-killing substances they produce. *Salmonella typhi* has found a way to speed up the process to aid its invasion plan. It carries genes that produce proteins with the ability to activate phagocyte caspases, inducing premature apoptosis, killing the cell and facilitating the release of more copies of bacteria. *Salmonella typhi* has not simply acquired the ability to evade phagocytic destruction, the usual fate of bacteria engulfed by these cells; it is a veritable Trojan horse, an enemy within, thriving inside the very cells that should be orchestrating its destruction, instructing them to commit suicide.

Although typhoid fever is associated with filth and squalor, the well-to-do were frequent casualties of the disease. Wealthy people with access to good sanitation and nutrition were more likely to escape exposure to gastrointestinal pathogens as children, and thus had less of a chance to develop immunity to typhoid fever early in life. Exposure to *Salmonella typhi* for the first time as adults resulted in more severe cases of the disease. Some people believed that typhoid fever exploited the "flawed constitution of immigrants" by using them as carriers of the disease to wipe out the "better class." Perhaps the most famous upper-class victim of typhoid was Prince Albert of England, the beloved husband of Queen Victoria. He became ill with it in 1861, after returning from a trip that was prompted by the unseemly behavior of his son,

Albert Edward (the future King Edward VII), who had been carrying on an affair with an Irish actress. Prince Albert lingered for about a month before dying, apparently from an intestinal perforation. The death of Victoria's young husband to a disease associated with filth shocked Great Britain. The prince consort, who had been an advocate of science and technology and who had sponsored the first World's Fair–the Great Exposition of 1851 in London, housed in the colossal Crystal Palace (an architectural wonder constructed of nearly 200 miles of glass, showcasing the most modern farm machinery and industrialized goods)–had succumbed to a disease that thrives in conditions produced by primitive sanitary habits.

Prince Albert was not the last member of the royal family to fall victim to typhoid fever. Albert Edward nearly died of a severe bout of it in 1871; and Albert Edward's son Eddy (Prince Albert Victor Christian Edward) contracted the disease two decades later and died at age 28. Eddy had acquired his father's love of gambling and drinking, a similar penchant for actresses and commoners, and the same susceptibility to *Salmonella typhi*. Eddy is known today because of his alleged association with the Jack the Ripper case. For the past 20 or 30 years he has been regarded by some forensic historians and amateur sleuths as a prime suspect; however, he was probably too dim-witted to have outsmarted the police by committing the serial murders without getting caught.

With the emergence of improved water processing and other measures to control infectious disease, such as pasteurization, typhoid fever acquired by ingesting contaminated food and water from commercial and municipal sources began to decline. But, by the beginning of the 20th century, physicians and public health officials began to recognize a new source of outbreaks, one not previously appreciated as a problem: chronic *Salmonella typhi* carriers. Nearly 5 percent of people who recover from typhoid fever continue to harbor the microbe in their bodies without displaying outward signs of infection. Carriers shed large quantities of *Salmonella typhi* in their feces and urine, which can infect others when it is transmitted directly from person to person through poor hygiene. About 25 percent of carriers have no history of typhoid fever at all, somehow managing to escape the disease despite exposure to the bacterium. The major reservoir for *Salmonella typhi* in carriers is the gall bladder, which the bacterium finds via a serpentine route that takes it through the twists and turns of the small intestine and bile ducts. Its ability to reach safe haven in the gall bladder is due to its acquisition of bile-neutralizing genes, located on one of its pathogenicity islands. Carriers release hundreds of millions of *Salmonella typhi* into the small intestine–and ultimately into the stools–when fatty meals are ingested, causing the gall bladder to contract and releasing the bacteria-laden bile into the small intestine.

Even today, there are still a few thousand carriers in the United States, mainly elderly women who acquired the bacterium in the early 20th century. For reasons that are not understood, women become chronic carriers of *Salmonella typhi* more frequently than men. Women also develop gallstones more often than men, although there is no obvious connection between these two phenomena.

The most notorious carrier of the typhoid bacillus was Mary Mallon, better known as Typhoid Mary, an Irish immigrant who came to the United States as a teenager at the beginning of the 20th century. An excellent cook, she was periodically employed by a number of well-to-do families who divided their time between their fashionable Park Avenue or Fifth Avenue mansions in New York City and their country homes on Long Island. She was known for her desserts, one of which was home-made ice cream with freshly sliced peaches—a perfect temporary refuge for *Salmonella typhi*, which survives cold temperatures quite nicely. Mary Mallon's downfall began in 1906 when six out of eleven members of the Charles Henry Warren household came down with typhoid fever. The family had rented a house from Mr. and Mrs. George Thompson on Oyster Bay, Long Island, a favorite country retreat for Manhattan's wealthy; President Teddy Roosevelt had a home there. Because of the association of typhoid and filth, an outbreak of the disease in a wealthy household had the potential to become a social nightmare. The Thompsons hired George Soper, a Columbia University–trained epidemiologist and specialist in sanitation engineering, to pinpoint the source of the outbreak.

After ruling out the usual causes–the water supply, local clams, and milk–Soper began to suspect Mary. She was the only new member of the household, and the typhoid outbreak occurred soon after she came to cook for the Warrens. After interviewing her previous employers, Soper discovered that from 1900 until the Oyster Bay outbreak, seven out of eight families for whom Mary had cooked came down with typhoid fever. In all, 22 people were affected, and there was one fatality. Soper had learned about healthy carriers of typhoid from European doctors and began to develop a case against Mary. After she returned to New York City, Soper confronted her. Angered by his veiled accusation of poor personal hygiene, Mary threw Soper out of her apartment, twice. He eventually returned with officers from the New York City Health Department, who forcibly took the belligerent cook into custody (not before she tried to chase them off with a pan). Mary felt persecuted. She could not comprehend the possibility of being a source of such a serious infection. Her health was excellent, and she denied ever having had typhoid fever. However, after being taken into custody, her urine and stool were examined and found to be teeming with *Salmonella typhi*.

There was no treatment available for carriers; effective antibiotics that could kill the hiding bacteria were nearly a half century away. Laxatives and a chemical called hexamethylenamine were tried as treatments by some doctors, to no avail. The removal of the gall bladder, a routine operation today, was too risky a procedure at the time to consider as a treatment for an asymptomatic carrier, and it was not completely effective, since other reservoirs of typhoid bacilli existed in the gut of carriers. Although some food handlers who carried the bacterium had been allowed return to society, as long as they promised not to work with food, Mary Mallon was viewed as a risk because of her lack of cooperation with health officials and her tumultuous temper. There may have been as many as ten thousand carriers in the United States at the time, and forced isolation of all of them would have been impractical and expensive, if not downright cruel. But Mary was a special case. Health officials recommended isolation. She was confined for the next three years to a bungalow on the grounds of a hospital on North Brother Island in New York City's

East River (where many active tuberculosis cases were housed), with only the company of a small dog, other patients, and a few staff members. Only after hiring an attorney, who filed a writ of habeus corpus on her behalf, was she ultimately released. The money for attorney fees apparently came from a wealthy sympathizer who was moved by Mary's lonely plight. She was 40 years old when she was finally released. The presiding judge warned her not return to professional cooking. She was soon forgotten by public health officials as she blended in with the other millions trying to make a living in the crowded city.

Mary tried other jobs, including working for a while in a laundry. But options were limited for an immigrant Irishwoman with little formal education. Her most lucrative career option was cooking, and eventually she returned to it under an assumed name, Mrs. Brown. But her decision proved costly. Unable to fully comprehend the need for extreme care in personal hygiene, she transmitted her hidden reservoir of *Salmonella typhi* once more, this time to patients and staff at the Sloan Maternity Hospital in New York City. She was apprehended again when inspectors from the Board of Health investigated an outbreak of 25 cases of typhoid fever at the hospital after two people died there.

She was denounced relentlessly in editorials, lampooned in cartoons, and given the venomous nickname that has endured for nearly a century. A cartoon in the newspaper *New York American* showed a caricature of Mary cracking a skull open into a skillet, like a fried egg. Anti-Irish sentiment may have played some role in the public loathing she had to endure. The Irish were considered one of the more unsavory immigrant groups. Many of them lived in extremely squalid conditions and suffered from poor health; tuberculosis was rampant, and infant mortality was high. For a time, the Irish in the United States had a lower life expectancy than their relatives who stayed behind in Ireland, becoming the only immigrant group in America with this reversal of health fortune. The nickname, "Typhoid Mary," was the creation of an imaginative editor from a newspaper owned by William Randolph Hearst, the same chain that had helped start the Spanish-American War in 1898 with biased, inflammatory reporting of the Cuban fight for independence (Hearst told one reporter, "You furnish the pictures, I'll furnish the war," which sums up the sensationalist approach that had been dubbed "yellow journalism" earlier in that decade). "Typhoid Mary" became the synonym for a 20th century leper.

Even the progressive journal *Scientific American* joined in the attack, accusing her in an editorial in 1915 of "competing with the Wandering Jew in scattering destruction in her path."

Vilified and hated, Mary Mallon was sent back to North Brother Island, where she was confined for the 23 remaining years of her life. Resigned to her fate, she spent her days helping tubercular patients and worked for a time—ironically enough—in the hospital's bacteriology lab.

Vaccines became the most important medical intervention against infectious diseases during the 19th century, and the one developed against typhoid fever made history by being the first that used heat-killed microbes to fool the immune system. Other vaccines developed in the late 19th century used live, but weakened microbes to stimulate immunity. Edward Jenner is usually credited as the founding father of vaccination, with his discovery in the late 18th century that intentional inoculation with cowpox prevented smallpox. However, some form of protective vaccination against smallpox had already been practiced for centuries in Asia and Africa (more on smallpox vaccines in chapter 13). Regardless of who should get the credit for introducing the practice of vaccination, it was not until the emergence of microbiology as a distinct discipline in the second half of the 19th century that a rational scientific basis for vaccine development began to emerge. The catalyst for this very important moment in human history was Louis Pasteur (1822-1895), who, in the 1870s and '80s, became the first person to develop vaccines using microbes cultivated in the laboratory.

Pasteur began his career as a promising chemist. His first major experiment—from which, it could be argued, all his great microbiological and vaccine discoveries followed by logical progression—was the resolution of an apparently arcane problem concerning tartaric acid, an organic chemical deposit found in wine casks. In the early 1800s, chemists discovered that tartaric acid came in two forms: isomers-tartaric acid (extracted from grapes and wine), and paratartrate (which could be synthesized in the laboratory). Each was identical in chemical composition and reactiveness. There was one difference: if polarized light was shined on a solution of tartaric acid, the beam was rotated to the right, while paratartrate failed to rotate light at all–an anomaly that perplexed the greatest chemists of the day. Pasteur came up with the simple idea of synthesizing paratartrate and sorting individual crystals by hand. Using a handheld lens, he observed that individual crystals fell into two groups that were mirror images of each other, and which were formed in precisely equal quantities. Solutions of each of these rotated light in the opposite direction from the other–the right-handed ones to the right, and the left-handed ones to the left; together, they canceled out each other's light-deviating effects. The crystal that deviated light to the right turned out to be pure tartaric acid. Pasteur had discovered a fundamental principle of organic chemistry: carbon-based molecules randomly form left- and right-handed isomers when synthesized in a test tube. The experimental results were so clear, and the implications so monumental to chemists of the day, that it was repeated step by step in front of a prominent scientist, Jean-Baptiste Biot, who became exuberant after witnessing the beautiful experiment firsthand. The giant intellectual leap for Pasteur, though, was his sharp observation that whereas organic compounds synthesized in a test tube were optically neutral, the same product extracted from living things showed optical asymmetry–that is, they rotated polarized light, and always to the right.

Optical asymmetry, Pasteur observed, distinguished the living and mineral worlds. From this Pasteur made the experimental observation that changed his career and the world of medicine. He found that when a solution of optically inert paratartaric acid was left in fermenting liquid for a period of time, only left-handed tartaric acid remained. Fermenting liquid selectively decomposed right-handed tartaric acid, a finding that led Pasteur to deduce that a living organism must be involved in the production of alcohol.

Despite its being such an important commercial and social enterprise, little was known about the fermentation of sugar into alcohol, even though people had been making wine and beer for thousands of years. Fermentation was believed to be an intrinsic chemical property of grains and fruits. In 1680, the inventor of the microscope Anton van Leeuwenhoek examined brewer's yeast and discovered that this critical fermentation starter, considered a secret key ingredient in the production of alcohol, contained round microscopic globules the function of which was unknown. These globules were assumed to be products of the fermentation process; thus, their presence was ignored, and their significance remained a mystery.

Pasteur's interest in fermentation took a practical turn in 1856, after his appointment to a prominent position as dean of the Faculty of Science at the University of Lille, when he was approached by a beer manufacturer with a problem. Vats of beer were turning sour and had to be thrown away, and Pasteur was asked to investigate the cause. He took up the challenge and began by first studying normal fermentation. He showed that the round globules in fermenting liquid first observed by Leeuwenhoek were living creatures, yeasts, and that they were critical for alcohol synthesis, demonstrating that when they were heated and destroyed, the liquid would not ferment. Thus, within a few years of his revelations about tartaric acid, Pasteur had discovered that the synthesis of alcohol was caused by microbes.

However, more important for beer manufacturers, Pasteur also showed that a second organism, a contaminating anaerobic bacterium, was responsible for causing beer to turn sour and rot. Like Richard Feynman, one of the great 20th century physicists, Pasteur was also a public experimenter par excellence (although to some, he was a relentless self-promoter). Feynman showed at a public congressional hearing that the explosion of the space shuttle *Discovery* in 1986 had been due to the loss of elasticity of an "O" ring seal in cold temperatures; he did this by simply dipping it in ice water and breaking it with bare hands in front of members of Congress and the press. More dramatic, and seemingly magical (because of the limitations of 19th-century scientific knowledge), Pasteur was able to accurately predict in public–by examining brews for the microscopically small contaminating culprit–which vats of beer would be ruined. Although those who witnessed the demonstration enjoyed the apparent drama of his feat, Pasteur had little doubt about the outcome. He did not conduct public demonstrations unless the experiment had already worked to his satisfaction over and over again in the laboratory.

After these fundamental discoveries, Pasteur showed that putrefaction could be prevented by heating the vats to a temperature high enough to kill the contaminating bacterium, but not so high that fermentation or taste would be adversely affected. This led to the process of heat sterilization to kill putrefying or disease-causing bacteria, called pasteurization in honor of Pasteur's work. Later, Pasteur proved that putrefaction of wine and the souring of milk were also caused by contaminating bacteria, and that they could also be prevented by this method.

Many scientists at the time believed that bacteria responsible for putrefaction emerged by spontaneous generation in fermenting liquids, milk, and other organic substances. Pasteur did not believe in spontaneous generation. He claimed that contaminating bacteria came from external sources, and conducted the now-famous "curved flask experiment" to prove it. Pasteur poured a liquid brew of sugar and albumin (the protein in egg white) into two flasks, one of which had a neck bent into swan's-neck curve; he heat-sterilized them, then quickly sealed them. No bacterial growth appeared in the sterilized liquid of either flask. He then exposed the sterile flasks to the air by breaking the seals; the openings were similar in size, yet microbial growth only appeared in the flask that opened directly to the air; the swan's-neck flask did not allow particles from the air to enter and contaminate the flask, because its long, curved neck blocked their entry. Pasteur reasoned that microbes were carried by air and were trapped in the curved flask's twisted neck-which kept them from reaching the liquid brew. Thus, the experiment demonstrated that bacterial growth originated from an external source.

Pasteur also helped the vinegar industry with his microbial discoveries. Vinegar is acetic acid, which is synthesized from alcohol. Pasteur found that the vinegar starter used for centuries by manufacturers carried a microbe, *Mycoderma aceti*, which uses atmospheric oxygen to convert alcohol to acetic acid, the chemical that provides vinegar with its acrid taste and pickling properties. Previously, it was believed that fermenting sugars exposed to air caused acetic acid to spontaneously develop from alcohol. The process was unpredictable. By showing manufacturers how to control the art of vinegar making through cultivating *Mycoderma aceti* and regulating the amount of microbe added to vinegar vats, the process could be standardized, producing a more consistent product.

Pasteur, through his work with beer, wine, milk, and vinegar, almost single-handedly helped save some of France's most important industries from financial ruin. In the course of this research demonstrating the ubiquity of microbes in the natural world and their role in the production of important manufactured goods, both positive and negative, he began to ponder the possibility that these microscopic creatures might also be responsible for disease. This concept, which was revolutionary in his time and led to a century and a half of discovery that continues to this day, was given credence by Pasteur's investigation, in 1865, of diseases that were destroying silkworms in the principal silk-producing regions of France. Silk, harvested from silkworm cocoons, is a soft yet durable fiber coated with sericin, a protein. The harvesting of silk had been practiced in China beginning in about 2640 BCE, and had become an important industry in Europe during the Renaissance after entrepreneurs imported silkworms from the Far East. Silkworm cocoons contain about a half mile of tightly woven silk that workers could unwind after throwing them in very hot water, a process that killed the developing silkworm (and provided a tasty caterpillar snack for some workers). In the early 1860s, silk production was failing miserably throughout Europe because of the loss of cocoons caused by pébrine, a parasitic protozoan disease that strikes insects. Only Japan, still isolated from the rest of the world at that time, did not suffer losses. Close examination of affected silkworms showed blotches on skin and internal organs, resembling tiny flecks of pepper, which under a microscope contained tiny oval corpuscles–these were the protozoans, *Botrytis bassiana*. Although Pasteur did not find a cure for the infection, he showed silk manufacturers how to identify the problem using microscopic examination and instructed them to isolate and destroy all infected silkworms. In the course of investigating pébrine, Pasteur also discovered a second disease, flacherie, which is caused by a vibrio bacterium that spreads by the silkworm equivalent of the fecal-oral route; infected worms contaminate others through fecal contamination of their only source of food, mulberry leaves. Pasteur taught manufacturers how to recognize infected silkworms by their irregular movement climbing trees. Destroying infected worms helped control the problem.

Pasteur's work with infectious diseases in silkworms was his gateway for establishing that microbes were also responsible for human diseases. This was more than a passing interest. Between 1859 and 1865, three of his young daughters died before reaching puberty, two of them from typhoid fever. Pasteur strongly suspected that airborne microbes were responsible for many human diseases, and noted the resemblance of some infections, such as gangrene, to putrefaction.

Like Koch, Pasteur's foray into human medical microbiology took an important turn with an investigation into anthrax, which was devastating sheep all over France and infecting farmers as well. In 1876, Koch isolated the anthrax bacillus, but experiments by Pasteur took the discovery a step farther. He showed that there were two forms of the microbe: the live infectious bacillus itself, and also corpuscles or spores, which were remarkably resistant to drying and heat, processes that killed the infectious form. Spores can turn into infectious bacteria when exposed to moisture and nutrients. Unlike the bacillus itself, the stable spore form of the bacterium can be turned into a bioweapon. Spores were commonly found by Pasteur in the soil of farms affected by anthrax, especially areas where carcasses of anthrax infected animals were buried. Pasteur deduced that spores were being brought to the surface by earthworms–creatures that Aristotle referred to as "intestines of the soil"–and thus enabling anthrax to infect other animals. He showed this experimentally by finding a large number of anthrax spores in the intestines of earthworms that fed on contaminated soil. The natural turning and aeration of soil by earthworms, so important for the health of a cultivated field, was also responsible for helping spread the anthrax bacillus to grazing farm animals and farmers. Pasteur recommended that the carcasses of dead animals be burned, or at the very least, buried far from feeding pastures.

While pursuing his anthrax research, Pasteur observed that animals that survived infection from experimentally induced anthrax became resistant to natural anthrax. Pasteur began to experiment with the growth conditions of anthrax bacillus in the laboratory to see if a strain could be developed that would give animals a sub-lethal infection while providing protection against the deadly natural infection. Through meticulous trial and error, like a chef perfecting a new recipe, Pasteur found a formula for developing a strain of weakened anthrax that stimulated immunity but did not cause disease. This was not Pasteur's first foray into vaccine development. A few years before his anthrax studies, Pasteur identified the bacterium that causes chicken cholera, a deadly chicken infection unrelated to human cholera. During a very hot summer, some chicken cholera bacillus stocks were damaged from exposure to higher-than-normal ambient temperature. The damaged bacillus failed to kill chickens. However, when test animals were challenged again with fresh strains of cholera bacillus, which should have killed them, they remained unscathed. Pasteur recognized that he had serendipitously stumbled on a phenomenon akin to Jenner's cowpox treatment to prevent smallpox. With Pasteur's astute powers of observation, an excessively hot Parisian summer and a few resilient chickens helped launch the science of vaccine development.

Through precise laboratory manipulation and trial-and-error, Pasteur had caused highly virulent anthrax bacillus to "de-evolve" into a viable yet non-pathogenic strain. Unknown to Pasteur and other scientists prior to the DNA era, virulent organisms grown under experimentally manipulated stressful conditions (such as unnaturally high temperatures) will sometimes shed or mutate genes for virulence and revert back to a strain resembling a non-virulent ancestor. The same virulence genes that help a dangerous microbe adapt and thrive in infected hosts can sometimes be a burden in artificial laboratory settings, and are often jettisoned to achieve maximal growth in a test tube or petri dish-after all, growing in the safe confines of a laboratory incubator makes genes that provide bacterial strains with the ability to survive in a mammalian gall bladder or to pass through the acid-rich environment in the stomach acid a bit obsolete.

On May 5, 1881, Pasteur administered his attenuated anthrax bacillus to sheep at a small farm in front of a large crowd of dignitaries and agriculture officials; unvaccinated sheep served as controls. A great deal of attention was paid to the experiment because anthrax had taken a very heavy toll on the farming industry that year. After two weeks, both the vaccinated and control sheep were injected with a highly virulent strain of anthrax. A month later, 21 of 25 control sheep had died of anthrax, whereas all but one in the vaccinated group survived. The experiment was so successful that farm animals were immediately inoculated by the thousands with the Pasteur vaccine. Within a few years, the mortality of farm animals from anthrax fell from 9 percent to less than 1 percent.

Pasteur soon conducted another public science demonstration while working on a vaccine against rabies–a universally fatal brain inflammation caused by a virus transmitted by the saliva of infected mammals and by contact with infected brain and spinal cord. It was a serious problem in rural France and other parts of the world. The illness still conjures fear, and an almost irrational dislike for the wild animals that sometimes carry the disease, such as bats and raccoons. During the 19th century, before animal vaccinations became available for pets and farm animals, rabies was transmitted primarily by dogs. The development of a rabies vaccine posed a technical challenge to Pasteur, since the causative agent could not be identified without the powerful microscopic techniques needed to examine viruses visually (which had not yet been invented). However, Pasteur and other microbiologists knew of the microbes' existence and realized that they were smaller than bacteria–when infectious material was passed through a filter small enough to capture bacteria, the filtrate still transmitted disease in test animals. Through painstaking trial and error involving repeated passage from one experimental animal to another, sometimes crossing species, and by drying the infectious agent, Pasteur eventually was able to obtain an attenuated strain that protected dogs from developing rabies.

In 1885, while still experimenting with his rabies vaccine, Pasteur was visited by a grocer and a nine-year-old boy, Joseph Meister, who were both attacked by the same rabid dog. The grocer's injuries were superficial, but the boy had been viciously attacked, leaving him with severe injuries. Observers had little doubt that his death from rabies was imminent. Pasteur, without having had the opportunity to test his vaccine in humans, administered the serial inoculations needed to induce rabies immunity. The vaccine worked, and Joseph Meister became to first person to survive a rabies attack. Within a year, hundreds of people flocked to Pasteur's laboratory in Paris for treatment, adding to his fame.

Pasteur believed that successful vaccines could only be made by using live microbes that were attenuated, reduced in strength in some manner, but for once he was wrong. Under some circumstances, killed microbes can tickle the immune system to generate protective immunity, and a vaccine against typhoid fever was the first example of this phenomenon. In 1897, Sir Almroth Wright (1861-1947), a British pathologist and microbiologist (who was later a mentor to Alexander Fleming), developed an effective typhoid fever vaccine using heat-killed *Salmonella typhi*, as did German microbiologists Richard Pfeiffer (1858-1945) and Wilhelm Kolle (1868-1935) in 1896.

Wright was professor of bacteriology at the Army Medical School at Netley. His main preoccupation was to develop a typhoid fever vaccine that would protect the British military, who were spread out all over Britain's colonies around the world and were suffering heavy losses from the disease. Thousands of British soldiers died of typhoid fever while fighting in the Crimean War (1853-56), and thousands more were dying while serving in India. Using himself as a guinea pig, Wright showed that treatment with heat-killed *Salmonella typhi* produced a substance in his serum that inhibited the bacterium's growth (it was an antibody that blocked the typhoid bacillus). He tested the vaccine on 2,835 volunteer soldiers, with encouraging results. However, even though heat-killed bacteria were used, the vaccine still caused side effects, tempering enthusiasm for its widespread use. The positive findings that Wright obtained in his preliminary experiments on human subjects were also questioned by Karl Pearson, a brilliant mathematician and one of the founders of statistical analysis, and by physicians swayed by Pasteur's belief that only live organisms could induce immunity.

Wright wanted to vaccinate all the British soldiers before they left to fight the Boer War in South Africa (1899-1902), but the military command blocked his efforts. Ultimately, only a small fraction of soldiers, about 14,000 in all, volunteered for treatment. Winston Churchill himself refused the treatment, relying on the "trust in his own health" to avoid sickness; mocking the medical profession, he asked them to develop a vaccine against bullets rather than one that made people sick. Of the 556,653 who served in the Boer War–the British empire's Vietnam–57,684 contracted typhoid fever, killing 8,225, compared with 7,582 who died in actual combat. There was a significant reduction in the death rate from typhoid fever among the soldiers who had been vaccinated. Thus, thousands of men died because of inaction by an immovable military bureaucracy and lack of support from the medical establishment. In a conflict that was called the last "gentleman's war," the majority of deaths did not occur on the battlefield in the pursuit of glory under the banner of "duty, honor, empire" but in army hospital cots due to perforated bowels caused by *Salmonella typhi*.

Resistance to vaccination was widespread in England, and not just limited to military bureaucrats. On one occasion, opponents dumped a shipment of typhoid vaccine overboard in Southhampton. Opposition stemmed from the serious and occasionally deadly side effects of smallpox inoculation. The earliest forms of vaccination against smallpox usually resulted in a controlled, mild case of smallpox, but were associated with complications and a substantial mortality rate. The Jenner vaccine was much safer, but on rare occasions it too could lead to serious complications, especially in the young, including death (see chapter 13).

Wright did not give in to the strong anti-vaccination opposition. He continued to improve and test his typhoid fever vaccine until the results could no longer be questioned by reasonable people. In an army test conducted on British troops in India in 1911, only 26 of 100,000 vaccinated soldiers came down with typhoid fever, compared with 303 of 100,000 who were not vaccinated–a 90 percent decrease. By World War I, Wright had persuaded the British military to allow large-scale vaccination of the troops. Although it was not mandatory, most volunteered for treatment. The typhoid fever vaccine probably saved tens of thousands of lives in that war. Only 20,000 cases and a 1,000 deaths from the disease occurred, the fewest in the history of armed conflict since the emergence of *Salmonella typhi* as a human pathogen.

Despite Wright's discovery of the first heat-killed vaccine, his reputation was tarnished by his radical views on antimicrobial drugs. It was his belief that drugs could not destroy bacteria once they had established an infection in the body. In his mind, treatment with drugs to counter microbes would adversely affect the body's ability to fight infections by impairing its immune response. Wright believed that boosting the body's natural immunity was the only reasonable treatment for infections. He was the model for the character Sir Colenso Ridgeon in *The Doctor's Dilemma*, a play written by his friend George Bernard Shaw. Wright's views on fighting infections are explicitly described in the play: "drugs can only repress symptoms, they cannot eradicate disease. The true remedy for all diseases is nature's remedy." Shaw goes on to declare in the play, "Stimulate the phagocytes. Drugs are a delusion."

Wright also disagreed with the practice of applying antiseptics directly to wounds claiming that it killed phagocytes, and in so doing was harmful to the patient.

Wright's beliefs were shared for a time, paradoxically, by his protégée Alexander Fleming (1881-1955), the discoverer of penicillin. Fleming worked in the immunology institute directed by Wright. He had earned an excellent reputation, and a comfortable living as well, as a result of his surgical skills. He was one of the most able practitioners in the use of salvarasan, the anti-syphilitic drug developed by Paul Ehrlich, which had to be carefully administered by intravenous infusion–leakage of the drug into the subcutaneous space destroyed muscle and fat tissue. In 1928, Fleming accidentally stumbled upon a mold that contaminated a culture of *Staphylococcus* bacteria he was cultivating on an agar-filled Petri dish. The contaminating mold spores probably drifted into Fleming's lab from the floor below where a colleague was growing molds for a study on their effect on asthma. The mold was *Penicillium notatum,* and it produced a substance that diffused through the agar, wiping out all *Staphylococci* in its path.

Fleming spent a couple of years testing crude extracts of *Penicillium* for antibacterial activity. He envisioned its use as a topical antibiotic. However, he was never able to extract a sufficient quantity for the large-scale animal studies needed to prove an effect against severe bacterial infections, nor was he able to purify the antibacterial factor, which was necessary before it could be considered for human use. In fact, Wright discouraged Fleming from pursuing testing in animals, viewing animal experimentation as not relevant to humans. Although Fleming's inability to purify penicillin was an understandable failing, considering that he was a physician and not a chemist, he should have followed up on his extraordinarily valuable observation by collaborating with someone who was well-trained in chemistry. However, he was thwarted by Wright who refused to hire a chemist for the institute. "There is not enough of the humanist in chemists to make them suitable colleagues," he claimed. Fleming had to make do with his own limited ability in the field. He also received assistance from a recent medical graduate who had an interest in chemistry, Frederick Ridley. Ridley made a gallant attempt at isolating penicillin, but he was too inexperienced.

For all his good fortune and insight in discovering the antibacterial properties of penicillin, Fleming never realized the full therapeutic potential of his finding. Considering that penicillin eventually became the wonder drug of the world, Fleming's scientific approach was, to put it kindly, dispassionate. Wright's negative view of antibacterial drugs played some role in Fleming's ambivalence. Eventually, he abandoned the penicillin project.

Penicillin was successfully isolated by Ernst Chain (1901-1979) and Howard Florey (1898-1968) in 1938 after Chain read Fleming's original 1929 paper and immediately recognized the substance's tremendous potential. The original mold isolated by Fleming, which Chain used in his initial extraction experiments, turned out to be a poor producer of penicillin. There was too little available for adequate testing and treatment. Albert Alexander, the first recipient of penicillin, lingered in hospital for months with a severe infection but improved dramatically after he was given the antibiotic. But his infection was so widespread that he needed long-term therapy. There was simply not enough penicillin available to cure the infection and he eventually died. In 1940, Florey and his chief biochemist, Norman Heatley (1911-2004) moved from Oxford, England, to Peoria, Illinois, fearing that the continual Blitzkreig attacks on England by the German Air Force would destroy the facility. It was also becoming increasingly difficult to obtain supplies for their work in England during the war. The inventive Heatley, the often forgotten fourth major player in the penicillin story (along with Fleming, Chain and Florey), had to convert hospital bedpans into culture bins for growing his molds in large quantities.

After the move, a national search for a more productive penicillin-synthesizing mold was launched by U.S. Army officials, who, impressed with the early clinical trials, desperately wanted to use the drug to treat soldiers injured on World War II battlefields. There was also interest by the U.S. Department of Agriculture, who wanted to find new commercial markets for wheat and corn, which provided the extracts needed for large-scale growth of beneficial molds. People from all over the country began to send bread mold and fruit mold to the lab. A research assistant in the Peoria penicillin lab, Mary Hunt (dubbed "Moldy Mary"), was instructed to search local markets for moldy fruit. One day she brought back to the lab an attractive golden mold found on a rotting cantaloupe from a local market. It was *Penicillium chrysogeum,* which was found to produce two hundred times more penicillin than its weak cousin, enabling low-cost production of industrial amounts of the drug to begin. Penicillin was available for use by D-Day in 1944, just in time in time to save tens of thousands of severely injured soldiers before the Germans and Japanese surrendered a year later.

Fleming, Chain, and Florey shared a Nobel prize for their work, but Fleming was viewed by the public as the primary intellectual force behind the discovery of penicillin. Wright was a major factor in this historical misinterpretation. With the great success of penicillin, Wright began to realize the value of antibacterial drugs and wanted his institute to be recognized. Wright wrote a letter to the *London Times* claiming that Fleming deserved the credit for the discovery. Florey had always been reluctant to talk to the press about his group's work on penicillin because he feared that patients and doctors would flock to his lab for the drug; there was not enough to make it available to everyone.

For his radical views on antibacterial drugs and their negative influence on Fleming, which probably delayed the isolation of penicillin by several years, Wright was saddled with offensive nicknames concocted by his detractors: Sir Almost Right and Sir Always Wrong.

An individual's genetic makeup influences his or her susceptibility to typhoid fever. There is some evidence that people with blood group O are more resistant to *Salmonella typhi* while those with type B are more sensitive–and those with type-O blood are also more prone to become asymptomatic carriers, compared with other blood types.

Another genetic factor that may play a role in typhoid susceptibility is the HLA locus, a group of dozens of genes that support the T-lymphocyte branch of the immune system. HLA (which stands for <u>h</u>uman <u>l</u>eukocyte <u>a</u>ntigen) genes encode proteins that are found on the outer surface of nearly every cell in the human body. Those found on phagocytic white blood cells function as a sort of biochemical "platter"–they bind to fragments of foreign proteins and "serve" them to T-lymphocytes. The foreign protein fragments themselves are prepared in phagocytes after ingested pathogens, mainly viruses and other intracellular microbes, have been chemically degraded and processed. T-lymphocytes containing surface receptors with the best fit for the proteins served by HLA proteins are stimulated, triggering numerous immune responses that help destroy the foreign invader.

The HLA genes (and the proteins they encode) are the most variable in the human genome. There are so many different permutations that, with the exception of identical twins, it is difficult to find two individuals with an exact match of all the HLA proteins found on cells.[17] Variability improves immune-system responsiveness. Each HLA protein has the capacity to bind to a finite number of different antigens. In the absence of variation, the ability to stimulate the T-lymphocyte arm of the immune system would be limited to only a small number of pathogens. HLA diversity increases the chance that at least one of HLA protein subtype will be able to bind to an antigen derived from a wide range of infectious microbes for presentation to the T-lymphocyte arm of the immune system. Typhoid fever susceptibility seems to be influenced by HLA subtype.

HLA diversity is so important for maximizing immune function that it may influence mate selection by acting as a sexual attractant. Female mice are attracted to the urine of males who have different MHC genes (the mouse equivalent of HLA)–mating between the two increases the odds of having offspring with a more diverse MHC system, which, in turn, would increase their capacity to respond to infectious microbes. Although the mechanism of mate selection revolving around the MHC locus has not been established, it may operate through the olfactory system. In mammals, sensory neurons located in the nasal cavity, called the vomeronasal organ, respond to pheromones, chemicals produced by animals that affect a variety of behaviors in other members of the species, including sexual attraction. Some MHC proteins stimulate the vomeronasal organ.

A similar quest for HLA diversity in mate selection may also operate in humans. In one study, women who were asked to respond to the smells of sweaty T-shirts at around the time they were ovulating found those worn by men with HLA genes different from their own more appealing. Whether it is the pheromones in urine or sweat that female mammals find more enticing, the goal is to provide HLA diversity to offspring. Since people related to one another share HLA genes, the unconscious quest to provide HLA diversity to our offspring may be one of the factors that

---

[17]HLA proteins also mediate transplant rejection. Donors and recipients have to be as well matched as possible in order for a successful transplant to occur.

discourages breeding between close relatives, a factor disregarded by Queen Victoria when she married Albert, who was also her cousin. Perhaps if Albert had married someone else, his susceptibility to typhoid fever, which he apparently passed down to his son and grandson, might have been buffered by a more typhoid-resistant HLA subtype.

Work done by Gerald Pier at Harvard Medical School supports the idea that typhoid fever may have left its mark on the human genome by contributing to the evolution of cystic fibrosis. Recent findings from his lab show that CFTR (the cystic fibrosis transmembrane regulator encoded by the gene that is mutated in cystic fibrosis) is used by *Salmonella typhi* in its infectious life cycle. *Salmonella typhi* begins its invasion of the wall of the small intestine by binding to CFTR located on cells lining the small intestinal wall. The concentration of CFTR protein normally expressed on these cells is inadequate for this microbe's invasion plan. However, the bacterium has found a way to redirect CFTR traffic within the cells by injecting proteins–through a type III secretion system–that cause CFTR to accumulate or "cluster" in one portion of the target cell's membrane. Clustering of a target receptor facilitates bacterial binding; the more hooks that are present in a confined area, the stronger the attachment.

In causing target receptors to cluster, *Salmonella typhi* is exploiting a normal feature of membrane anatomy. Receptors and other proteins embedded in cell membranes are shuttled to specific regions of the membrane to create unique functional domains. Perhaps the best illustration of this phenomenon is the clustering of hundreds of different membrane proteins at synapses, the junctions formed between the end of one nerve cell (neuron), and the beginning of another. Communication between neurons at synapses forms the major means by which neurons transmit signals. By confining neuronal communication to specific membrane regions through protein clustering, signaling goes faster. This enables the brain and muscles to operate in a fraction of a second, facilitating activities such as sight-reading music at the piano, running from a predator, and making bat contact with a major league fastball.

Clustering of CFTR is just part of an extensive membrane alteration and reorganization plan initiated by proteins injected though the type III secretion system, all of which are designed to facilitate bacterial entry. Infected cells are, in essence, turned into cellular zombies, performing strange biochemical tasks according to *Salmonella typhi's* commands.

Some people are born with the ability to thwart *Salmonella typhi* by interfering with its invasion scheme. Approximately 3 to 4 percent of Caucasians have mutant *CFTR* genes coding for CFTR proteins that do not bend to the bacteriums' will. Either the protein fails to migrate to the surface membrane, or its capacity to act as a hook for the bacterium is impaired. In some cases, mutant *CFTR* genes fail to generate CFTR protein at all. Regardless of the biochemical and genetic mechanism, CFTR clustering is reduced when mutant *CFTR* genes are present, which reduces the ability of the typhoid bacilli to invade cells.

Mutant *CFTR* genes, which started to appear in the human genome as long ago as fifty thousand years, may improve the odds of surviving a *Salmonella typhi* attack. But with typhoid fever being such a rare disease in the United States and Europe today, protective *CFTR* gene mutations have become more of a burden, genetic relics from a different era. Carriers of these genes living in developed nations would gladly trade in these now-unnecessary mutations to rid them of what has become a far greater health concern. Parents who both carry mutant *CFTR* genes can transmit two copies to their offspring, giving them cystic fibrosis.

CHAPTER 10

The Salty Sweat Disease

In addition to the inherited traits that endow us with our particular visible characteristics (height, skin color, etc.) and elements of our personalities, we also possess hidden genetic lives. Buried within the chemical code of virtually everyone's genome are genes for latent traits that may manifest late in life or may only appear after we are exposed to specific environmental factors. Unbeknownst to us, we may carry genes for Alzheimer's disease or osteoarthritis, waiting for advancing age to creep up before their damaging effects on the brain and body emerge; or genes that make us susceptible to a severe reaction to some medications (such as penicillin) and foods (such as fava beans and shellfish); or genes that increase the risk of developing blood clots. But the effects of such genes will not be experienced unless the offending drug is inadvertently prescribed or the risky food consumed, or until an extensive period of forced immobility–sitting in one position for hours on a long plane trip, for example–causes a blood clot to develop in a leg vein, which can then break off, travel to the lungs, and cause sudden death.

For many individuals, the hidden genetic trait is a recessive gene for a serious inherited disorder that has no major effect on their own physiology but, should they have the misfortune of partnering with someone who has the same defect, will damage their children. Peruse a group of thirty people in a classroom, passing you by on a busy street, or sitting on a bus or train, and you will probably glimpse at least one person who is harboring a mutation in the *CFTR* gene that can transmit the fatal inherited disorder cystic fibrosis. There is nothing visible that will distinguish him or her from anyone else, other than having a small change in the *CFTR* gene that would require precise DNA analysis to identify. Most are unaware of their carrier status, and are blind to the legacy of disease that allowed this mutation to prosper throughout humanity. Because cystic fibrosis is an autosomal recessive disorder, carriers usually learn about their hidden genetic problem during prenatal testing or, worse, after discovering that their child has the disease. It is the most common fatal inherited disorder in Northern Europeans and their descendants; overall, about 3 to 4 percent of Caucasians carry the trait, whereas its frequency in blacks and Asians is very low. Caucasian carriers have roughly a 1-in-30 chance of having a partner who is also a carrier, with the result that about 1 in 900 Caucasian couples (1/30 x 1/30) who mate will both be carriers for cystic fibrosis, and their chance of conceiving a child with two copies of the abnormal gene, resulting in the disorder, would be 1 in 4. The upshot of these simple calculations and probability determinations is that 30,000 people in the United States and 70,000 throughout the world have full-blown cystic fibrosis.

The physiological problems underlying cystic fibrosis have been known for a number of years. The fundamental defect is an imbalance in salt and water reabsorption in several different tissues that causes concentrated or thickened body secretions, and a failure to clear certain microbes from the respiratory system. The consequences of these physiological mishaps range from the serious (but not life-threatening), such as male infertility, to problems that can be lethal: bowel obstruction in newborns, poor absorption of nutrients from the intestines (caused by a decrease in digestive juices secreted from the pancreas), and pulmonary failure.

Regulating the transport of salts (NOTE: the terms electrolyte and ion will be used interchangeably with salt) across the cell membrane is an essential physiological function that organisms must control in order for cells to work properly. Salt regulation, especially sodium chloride, the most abundant salt in the body, is intimately connected with cellular water. Water flows in and out of cells by osmosis (diffusion across cell membranes), primarily, but not exclusively, as a consequence of differences that exist in the concentration of sodium chloride inside the cell, compared with the concentration outside. Basically, if the salt concentration is higher outside the cell, water will flow out; if the concentration is higher inside, water flows in. Thus, without the proper balance of salts and water, cells would either shrivel up from dehydration or burst with excess water. When the sodium chloride level in the blood falls by about 10 to 15 percent, it can cause a state of mental confusion, coma, and, ultimately, death from brain cell swelling.

Salts also move across cell membranes; the natural direction of flow is from the side with a high salt concentration towards the side with a lower concentration. Scientists call this a concentration gradient. The movement stops when an equilibrium state is reached; that is, an equal concentration is achieved on both sides of the membrane. It is the same type of principle that guides the dispersal of milk when it is poured into a hot cup of black coffee-the milk will be distributed equally throughout coffee until a uniform color–a state of equilibrium–is reached. However, salts cannot simply cross cell membranes at will; the fatty portion of cell membranes is a barrier against the movement of water soluble substances, such as salts. Instead, the movement of various salts in and out of cells is regulated by a variety of membrane proteins-ion channels and pumps-through which salts move. Some ion channels are "gates"–they open and close in response to a variety of physiological states, such as the electrochemical current generated by activated neurons and muscle cells, an infusion of calcium ions, or an increase in the second messenger molecules cAMP and cGMP (cyclic AMP and GMP–cyclic nucleotides of adenosine and guanosine, respectively). Each of these phenomena can directly or indirectly alter the three-dimensional configuration of the channel, causing the pore through which ions move to open.

Some cells have specific needs for certain salts that cannot be satisfied by passive movement through ion gates, in which case, protein pumps are activated to drive ions into or out of cells, against a concentration gradient. For example, muscle and nerve cells need to establish a very high potassium concentration in order to maintain their electrical excitability (about 25 to 30 times higher than the level found in blood) and, concomitantly, a relatively low concentration of sodium, compared with the levels found outside the cell. Every electrochemical signal passing through a nerve or a stimulated muscle cell results in the influx of sodium ions and the efflux of potassium. This movement generates the electrical signal that drives all neuronal and muscle cell activity. In order to maintain high levels of potassium (and their electrical, excitable edge), neurons and muscle cells contain pumps that actively move potassium ions back inside. The same pump moves sodium ions out of the cell. Without a mechanism in place to constantly re-accumulate potassium and eliminate sodium, muscles would become useless slabs of meat after a few contractions, neuronal activity would cease, and the beating heart would weaken and fail in a few minutes instead of continuing to function over a period that is usually expected to last seventy or eighty years. However, like salmon swimming upstream, or animals walking uphill, biological systems that defy the laws of physics require energy in the form of ATP, which cells are able to synthesize from a variety of different chemical pathways encoded by dozens of genes. The chemical energy stored within a molecule of ATP can temporarily move bodies against the force of gravity and move salts against a concentration gradient.

Ion channels and pumps are also used to preserve body salts. In the sweat gland, for example, pumps reabsorb sodium chloride that is naturally excreted in sweat. People sweat as a mechanism to rid the body of excess heat; otherwise, our body temperatures would rise to dangerously high levels during intense physical activity on warm days. The body cools when sweat evaporates from the surface of the skin, which releases heat. Sodium chloride is not needed for heat dissipation. However, sweat glands are unable to secrete pure water; some salt loss is inevitable. Since excessive loss of salts can deplete the body of sodium chloride to dangerously low, even deadly levels (a condition called hyponatremia), evolution has provided us with pumps that reabsorb sodium chloride from the skin to minimize the loss. The pump is not perfect, which is why sweat is a little salty. However, the amount of salt excreted in sweat is small enough to be easily replaced by salt taken in by eating food and by drinking sports drinks and fruit juices, which contain sodium, potassium, chloride, and other electrolytes.

Hyponatremia can occur while taking part in strenuous sports events on very hot days, when participants sweat several pints of salty water and replace it by drinking plain water, which has only trace amounts of salt (distilled water has no salt). Occasionally, severe hyponatremia will claim the life of a marathon runner who drinks too much plain water during a race on a hot day. This happens more commonly to average runners than to the top competitive athletes. The former lose more sweat than the latter, because they're less conditioned and take much more time to complete a race. The huge market that has developed around sports drinks has a scientific rationale.

Dangerous hyponatremia also occurs with the club drug, Ecstasy. The drug increases body temperature, an effect made worse by all-night dancing and the body heat generated by hordes of similarly affected fellow clubbers. To help cool down, people on Ecstasy drink large volumes of cold water, which can dilute serum sodium chloride to such a low level, uncontrollable epileptic seizures may set in.

Different channels and pumps exist for different ions, with potassium, sodium, bicarbonate, and chloride being the most common. Some channels can accommodate more than one type of ion. The basic physiological problem in cystic fibrosis is a defect in the capacity for some cells to move chloride and bicarbonate ions across the cell membrane because of the inheritance of two copies of mutant *CFTR* genes. *CFTR* codes for an ATP- and cAMP-regulated chloride channel; when the ATP binds to the channel it opens the pore allowing ions to move through. Since the most frequent partner of chloride is the sodium ion, an abnormality in CFTR channel production affects the concentration of sodium chloride in various body fluids, and ultimately water, whose movement in and out of cells is primarily driven by the level of sodium chloride.

One of the tissues affected in cystic fibrosis is the sweat gland, where lack of normal CFTR leads to a failure in the reabsorption of sodium chloride and, consequently, the secretion of excessively salty sweat. This physiological mark forms the basis for a screening test to diagnose cystic fibrosis in infants, measuring salt content in the sweat formed on a patch of skin. Salty sweat does not usually pose a major health problem for most people, as long as the lost salt is replaced and their level of physical exertion on hot days is moderate, so as to prevent severe hyponatremia. However, in full-blown cystic fibrosis, excessive loss of salt can be dangerous if not recognized and treated. In 1953, Dr. Paul di Sant' Agnese attributed a sudden increase in deaths from cystic fibrosis to a severe heat wave that hit New York City that summer.

There is an old Northern European adage that seems to suggest that children with cystic fibrosis, and their dire prognosis, were recognized centuries ago by their distinctive sweat problem: "Woe to the child when kissed on the forehead tastes salty. He is bewitched and soon must die."

The first clue that a newborn child has cystic fibrosis is intestinal obstruction. Newborn babies produce a very dark stool called meconium. Fetuses under duress sometimes have a bowel movement *in utero* that is visible as meconium-stained amniotic fluid, after women in labor rupture the membrane that holds amniotic fluid in place. Meconium is very thick and gluey in newborns who have cystic fibrosis, almost tar-like in consistency, and it can obstruct the intestinal tract at its most narrow passages. This occurs in about 10 percent of babies born with cystic fibrosis. Surgery may be needed to relieve the obstruction. While heterozygotes for mutant *CFTR* may have prospered in evolution by protecting children against the ravages of secretory diarrhea, homozygous newborns suffer from the exact opposite dilemma.

The most common problem in older children with cystic fibrosis is pancreatic failure. The pancreas–a soft, elongated structure located behind the stomach–is the anonymous, behind-the-scenes organ of the digestive system. Everyone knows something about their stomach and intestines, and has some fundamental understanding of the function of these organs. However, few laymen, I would venture to guess, know much about the pancreas and the critical role that it plays in digestion. The pancreas is the principal producer of enzymes needed to break down the basic nutritional elements–proteins, carbohydrates, and lipids–into smaller components that can be absorbed by the small intestine into the blood.

Glandular cells located throughout the pancreas secrete digestive enzymes. The cells coalesce to form ducts into which the enzymes are secreted for passage into the small intestine via a tube called the pancreatic duct. In people who have cystic fibrosis, the enzyme-rich pancreatic secretions are thick and sludgy, which obstructs flow, essentially causing a backup like a clogged drain and producing disastrous consequences to the pancreas. The enzymes that were destined to help digest food after it passes into the small intestine essentially turn against their master and destroy the very portion of the pancreas involved in their synthesis.[18] Ultimately, after a few years of sustained damage, the pancreas is no longer able to secrete enough digestive enzymes to properly break down food particles, leading to malabsorption (poor absorption of nutrient material from the intestines). Malabsorption causes malnutrition and vitamin deficiencies, resulting in slow starvation even with an ample food intake, as happens to people who have tapeworms. Before treatment became available for this problem several decades ago (in the form of orally administered pancreatic enzymes, injectable vitamins, and intravenous nutritional support), malabsorption in childhood was the main cause of death in cystic fibrosis. It is only in the modern era that such patients have been able to routinely survive childhood, allowing them to live long enough to develop the terrible pulmonary disease that awaits them as young adults, a problem that has emerged as their chief nemesis.

People with cystic fibrosis are very susceptible to bacterial infections in the bronchioles of their respiratory airways, which leads to chronic inflammation, damage to those small airways and lung tissue, and, ultimately, pulmonary failure. Bacterial infections begin in childhood with severe sinus infections and attacks of bronchitis, and recur again and again. The airways' susceptibility to bacterial infections in cystic fibrosis is complex. Humans are

---

[18]The pancreas also produces insulin but uses a different set of cells for this purpose: beta cells. Insulin-producing beta cells are usually spared from destruction early in cystic fibrosis but may be affected as the disease progresses, causing diabetes.

exposed to a host of bacteria, viruses, and fungal spores when we inhale dust particles or the coughed or sneezed fomites expelled from another person's (or animal's) respiratory system. Humans are not very good at preventing infections from inhaled viruses, which is why most of us suffer frequent bouts with the common cold. However, the airways are quite efficient at preventing bacterial infections. Many of us living in the developed world can go through life and never experience an episode of bacterial bronchitis or pneumonia, despite our daily inhalation of bacteria-laden aerosolized particles. An important protective force against bacterial infection in the respiratory airways is mucus. Normal mucus forms a liquid blanket that coats the respiratory tree, trapping bacteria and particulate matter, protecting the airways and delicate lung tissue from infection. Mucus contains antibacterial substances, such as mucin, which can retard the growth of bacteria. Bacteria and dust particles of a certain size that are effectively trapped by mucus in the respiratory airways are expelled by coughing. We do this reflexively, coughing up barely visible chunks of this material, and usually swallowing the residue. The mucus produced in cystic fibrosis is thickened from a defect in the transport of sodium chloride across airway epithelial cells. The resulting alteration in salt concentration impairs the ability of mucin and other anti-bacterial substances to block bacterial growth.

Susceptibility to bacterial infection can also occur because ciliary action in the airways is impaired. Cilia are tiny hairlike projections that protrude from epithelial cells lining the bronchial walls. They are shortened versions of the long, whip-like flagella that propel sperm and motile bacteria. Ciliary movement produces a tiny wave of mucus that sweeps bacteria and dust particles away from the bronchial wall, preventing bacteria from penetrating into the submucosal layer where a focus of infection can be established. In people who have cystic fibrosis, thickening of the mucus impairs ciliary movement, as if the little hairs were trying to move in a dense, creamy soup instead of water.

As a consequence of thickened mucus and recurrent bacterial infection and inflammation, people with cystic fibrosis have a chronic cough and produce very thick, nasty-looking, infected phlegm. Although impaired salt and water balance is partly to blame for thick mucus, the increased viscosity is also a consequence of chronic infection and the inflammatory response it elicits, as well as the cause. As inflammatory cells in the airways die, their cell and nuclear membranes rupture, releasing DNA. Once uncoiled from its protein-bound packed state in the nucleus, free (or "naked") DNA turns an aqueous solution viscous. To observe this on your own, simply crush an onion with a mortar and pestle in a few tablespoons of water and add a few drops of liquid dishwasher soap. The DNA released from the crushed onion cells in the presence of a detergent–which essentially disrupts the proteins that keep the three-foot-long strand of DNA present in each cell in a coiled, condensed state–causes the solution to become very viscous. Killed bacteria also contribute to the pool of DNA found in the airways of people with cystic fibrosis. The realization that naked DNA released from its protein shackles was a major factor in the increased viscosity of the phlegm that accompanies cystic fibrosis led to a clever therapy; DNAse in aerosol form. DNAse is an enzyme that chops up complex DNA molecules into individual nucleotides or small strings of nucleotides, converting the genetic code of life into a pile of chemical rubble. When delivered to the airways of cystic fibrosis patients in an aerosol, DNAse breaks up naked DNA, thinning the viscous phlegm, and thus helping patients clear their respiratory tract of infectious waste.

Explaining the susceptibility to lung infection in cystic fibrosis by these purely mechanical processes is an oversimplification. If they were entirely accurate, one would expect patients to harbor dozens of different microbes in their airways. In fact, of the wide array of bacteria, viruses, and fungal spores constantly inhaled by humans, only a small handful are serious pathogens in cystic fibrosis, the most harmful of which is *Pseudomonas aeruginosa*. Approximately 90 percent of the people who have cystic fibrosis suffer from chronic *P. aeruginosa* infections. This bacterium elicits a potent inflammatory response that damages respiratory airways and lung tissue. After years of repeated bouts of infection and inflammation, lung function begins to deteriorate. Most patients spend their teenage years and twenties as "respiratory cripples," experiencing shortness of breath from the simplest exertion, even just walking to the bathroom (a similar fate awaits chronic smokers who go on to develop emphysema.).

Cystic fibrosis patients are treated with antibiotics, made to perform deep voluntary coughing to remove phlegm, and given DNAse aerosol to loosen airway secretions. These are very valuable tools that have doubled the average life span of these patients. However, death from pulmonary failure eventually occurs, usually by the time they reach the age of 25 to 30. Patients with a more mild form of cystic fibrosis, caused by less damaging *CFTR* gene mutations, are less debilitated by respiratory problems. The longevity record belongs to an American woman in her late 70s who inherited a mild form of the illness. There is a small contingent of runners in the New York City marathon with relatively mild forms of the disease who have maintained enough pulmonary function to jog the 26 marathon miles. But these are exceptions.

Until gene therapy or drug treatment can be perfected, the only chance of arresting the pulmonary problem in people with cystic fibrosis is a double lung transplant, which, if successful, can offer them a nearly normal life. The decision to undergo double lung transplantation is the game of roulette that cystic fibrosis patients and their families have to play. More than a third of these patients who receive a lung transplant die prematurely from the aftermath of the surgery, rejection of the transplant, or complications related to the immune-suppressant medication that has to be taken to prevent rejection. Survival is lower if surgery is performed when end-stage pulmonary failure has already occurred, so they can not wait until then to make the decision. Should they sacrifice the last few years of life for a chance at successful lung transplantation? If they can survive for a few more years without surgery, will gene therapy or some other major breakthrough emerge that would make lung transplantation obsolete? One unsuccessful transplant recipient was a Brown University student, Laura Rothenberg, a young woman who chronicled her life as a cystic fibrosis patient in a series of autobiographical audio pieces she called "My So-called Lungs," which were broadcast on National Public Radio.

Why do only a very small handful of organisms, in particular *P. aeruginosa*, cause chronic infection and inflammation in the airways of cystic fibrosis patients, despite the daily exposure to dozens of potential pathogens? One reason is a unique, intimate relationship that exists between *P. aeruginosa* and CFTR protein, a finding that had to await the discovery of the CFTR gene.

Identifying the mutation in a gene responsible for an inherited disorder in a genome consisting of three billion nucleotides is equivalent to finding an error of a single letter in a fifteen million-page manuscript. It would take a single reader nearly a century reading two average sized books a day to find the mistake. However, if there was information available indicating that the error was located somewhere in just a few books–thus narrowing the search and changing a Herculean task to a merely formidable one–it could be found in a matter of months. This is essentially what researchers analyzing the human genome try to do when confronted with the task of locating an unknown gene responsible for a disease: narrow the search from the entire three-billion-nucleotide genome to a more limited five- or ten-million-nucleotide patch in a discrete region on one of the 23 pairs of chromosomes. The principal strategy used to narrow the search for a mutant gene is called linkage analysis, a time-honored method that geneticists have applied to "map" or localize the position of genes to a specific chromosomal region. The beauty of linkage analysis lies in its broad applicability. It can be used to map genes for any trait in any organism, ranging from hair consistency in flies to neurological disorders in humans. Mathematics, more than biology, is its guiding principle, since linkage analysis is really an application of the law of probability combined with knowledge of how chromosomes and the genes they carry are transmitted to offspring.

The probability of inheriting two traits will differ, depending on whether or not the genes coding for the traits (specifically, the allelic versions or variants of the gene) are located on the same or different chromosomes. To give the simplest example: say there are two heterozygous traits in one parent that are located on separate chromosomes; call the genes "A" and "B." In gene "A" there are two variants called "A1" and "A2," while in gene "B" there are the "B1" and "B2" variants. Heterozygotes would have the gene structures "A1/A2" and "B1/B2." The probability of transmitting both "A2" and "B2" to an offspring is 1 in 4 (assuming the other parent has an A1/A1 and B1/B1 configuration), a product of the individual probabilities that each trait will be transmitted (½ multiplied by ½). However, if "A2" and "B2" are located close together on the same chromosome, the probability of transmitting both increases to ½ since they are being transferred to a germ cell as a single unit on one member of a chromosome pair (actually, the probability is a little less than ½ because genetic recombination can separate the two genes in some germ cells, but I will refrain from discussing this problem here, in order to make the concept easier to understand).

To apply this mathematical concept in gene mapping, researchers analyze hundreds and thousands of genetic markers in families with a genetic disorder. The markers are commercially available, and their chromosomal location has already been established. They then determine whether transmission of individual markers—and the illness—is more consistent with them being on the same chromosome (approximately 50 percent, assuming marker and trait are both heterozygous in only one parent) or different chromosomes (25 percent probability, under the same conditions). Once a marker is consistently found to be transmitted together with the illness, the gene must be located near the marker. Since the chromosomal position of the marker on that chromosome is already known, the mystery gene must be nearby—in essence, guilt by association.

Mapping and isolating genes by linkage analysis became a practical possibility with the advent of recombinant DNA technology. Before that, there were not enough genetic markers available to create a decent map. The first inherited disorder mapped by applying recombinant DNA technology was Huntington's disease (also called Huntington's chorea), a deadly neurological disorder that killed folk singer Woody Guthrie in 1967. The gene was first mapped to a region on chromosome 4 in 1983 and was finally isolated a decade later. Genes for familial Alzheimer's disease and many other conditions were also identified in this manner. However, the successful mapping and isolation of the cystic fibrosis gene was the top goal in the 1980s among genetics researchers, considering its high prevalence in the population and its terrible clinical consequences to children and young adults. The search for this gene attracted many of the best scientists in the field, motivated by the plight of families with the illness as well as by the idea of working on a "headline" disease, one with Nobel Prize potential, and it also attracted commercial interests. Many families with a child who had cystic fibrosis were opting not to have additional children, unwilling to take the risk of conceiving another baby burdened by such a tragic condition. Finding the gene, and the responsible mutations, would provide a genetic test that could be used for prenatal diagnosis in families at risk, allowing couples to choose a fetus free of cystic fibrosis. The potential market for prenatal testing was huge. Not only would couples known to carry the disease be interested, but a case could be made to test all Caucasian couples, even those without a family history, since most carriers are unaware that they harbor this genetic time bomb.

The race to find the cystic fibrosis gene created one of the first battles between the conflicting interests of the academic and commercial worlds. A company called Collaborative Research Inc. invested millions of dollars in developing markers that could be used for mapping the cystic fibrosis gene and other genetic disorders. However, they needed access to families with the illness, which could only be found in large clinical academic centers. One of the largest was in Toronto, where a team of investigators led by an idealistic researcher named Lap-Chee Tsui was collecting DNA from families affected by cystic fibrosis, for future linkage analysis. Other groups around the world–in Utah, Boston, and London–were doing the same. Tsui tested markers made available to him by Collaborative Research, and in 1985 he found one on chromosome 7 that was linked to the illness. Tsui wanted to provide all the available information to interested researchers at a conference, but was deterred by his Collaborative Research partners, who did not want him to provide the marker's precise chromosomal position. They feared that full disclosure would furnish competitors with information that would neutralize their own head start on the rest of the world. Full disclosure would essentially put everyone in the race on the chromosome 7 track. Providing other labs with critical unpublished scientific information is a necessary hazard of presenting at scientific meetings–necessary because it pushes the science forward and shortens the time before sick people can benefit from discovery, hazardous because a larger lab learning about new findings could end up scooping you. At least for publication in print media (usually a scientific journal), important findings can usually be kept somewhat confidential for a few months, the approximate time it takes for most scientific papers to weave their way through the publication process, providing the researcher in question with a decent advantage over the rest of the world (unless a reviewer spills the beans or uses the data for his or her own research, both of which are unethical practices). Eventually though, the chromosomal location of the cystic fibrosis marker was revealed, and Tsui's findings opened the gate: the race to isolate the cystic fibrosis gene had officially begun in earnest.

The first step in isolating a gene that has been mapped by linkage analysis is to narrow the region to a relatively small segment of genomic DNA, one that can be easily managed in bacteria and yeast, the laboratory workhorses of recombinant DNA (see chapters 14 and 15). Generally, a workable piece of DNA is about half a million nucleotides long and may contain about five genes, but it must be further subdivided into many smaller fragments for detailed analysis. In any case, it takes quite a bit of work to map a gene down to a half million nucleotides; initial linkage screens are only capable of identifying a marker within a range of 5 to 10 million nucleotides. Tsui's original marker was actually 15 million nucleotides away from the cystic fibrosis gene. To narrow the search, linkage analysis has to be painstakingly repeated using more and more markers in the defined region. In the 1980s, however, there were so few markers available that researchers who needed to pin down the locations of unknown genes more finely had to find new markers and create their own detailed maps, a routine but very time-consuming enterprise that took a couple of years. Fine linkage mapping is much easier today because, over the past few years, several million markers have been identified and analyzed by gene mappers in the course of the human genome project (see chapter 16). The difference between current genetic maps and those available in the 1980s is comparable to the information provided by a street map of every city in the United States versus a map that only shows the outlines of the states. Today, finding new markers for gene mapping projects requires only a few clicks on a computer mouse instead of a few years of extra lab work.

Once a gene has been mapped to a small chromosomal region, researchers can attempt its isolation. This is done by sifting through the final segment of DNA known to contain the gene and searching for hints of exons–the protein coding portion of genes, and the part most commonly affected in genetic disorders.

Since the completion of the human genome project, identifying coding elements in genomic DNA and determining the borders of any gene has become fairly simple, requiring a few mouse clicks while browsing through extensive DNA databases that have been posted on unrestricted Web sites. However, during the 1980s, identifying genes buried within a chunk of genomic DNA was very difficult, frequently taking several years of lab time to achieve. Recall from chapter 1 that only a fraction of the genome harbors genes, and that most of it is has no protein-coding function. Several clever strategies were developed to solve this problem, but the most effective one relied on taking the genomic DNA fragment isolated through linkage analysis and comparing it with sequences found in "cDNA libraries," which are derived almost entirely from exons, using a strategy called hybridization. Hybridization takes advantage of the affinity that one strand of DNA has for binding to another strand that shares a long stretch of common sequences. Exons located on a stretch of genomic DNA can be fished out using cDNA hooks (probes) that bind to regions of similarity. Thousands of different cDNA libraries have been constructed utilizing genetic material derived from cells and tissue from humans and other organisms, but only those derived from cells that express the unknown gene can be used. The choice of which cDNA library to use is fairly straightforward; if one were dealing with a neurological disease, for example, a brain cDNA library would be needed to fish out the unknown gene. In searching for the cystic fibrosis gene, choosing the correct cDNA library made all the difference.

Over a period of several years following Tsui's initial findings, the principal labs working on the cystic fibrosis gene had painstakingly narrowed the critical region on chromosome 7 to a fragment that only contained a handful of genes. In 1987, a London-based research team headed by Robert Williamson announced the identification of a cystic fibrosis candidate gene called *IRP*. It was big news. However, from the beginning, the results were questioned by other researchers in the field. From years of physiological studies, investigators suspected that the underlying defect in cystic fibrosis was in the transport of chloride. However, the protein coded by *IRP* had no apparent connection to chloride, or to any other ion for that matter. After the publication of their paper, there were rumors that Williamson's group was accumulating new data showing that *IRP* was indeed the wrong gene. The group had compared the *IRP* gene sequence in controls and patients with cystic fibrosis and could not find any significant differences. Most damaging was the finding of a few families in whom markers near *IRP* were not linked to the development of cystic fibrosis. Unless the genetics of cystic fibrosis was more complex than scientists believed, *IRP* could not be the gene.

Cystic fibrosis researchers were exasperated. Some felt that Williamson's group was not forthcoming with critical negative data. However, Williamson was so convinced initially that IRP was the long-sought cystic fibrosis gene, he was trying to come up with alternative explanations for his group's negative findings. Eventually, Williamson recognized that an error had been committed. It was a crushing defeat, made worse by their premature public announcement. Not only were they wrong, but the lab's focus on *IRP* had wasted precious time, effectively taking them out of the hunt for what turned out to be the true cystic fibrosis gene.

For the other major investigators, early skepticism about *IRP* resulted in an uninterrupted continuation of the laborious cloning and sequencing effort across the critical region of chromosome 7. Persistence eventually paid off for Tsui, who had joined forces with the research group headed by Francis Collins, Director of the National Human Genome Research Institute, who was then at the University of Michigan. The final critical piece of the genetic puzzle was based, fittingly, on the salty sweat phenomenon. Tsui and Collins reasoned correctly that a coding region should exist in their isolated genomic DNA fragment that would overlap with genetic material from a cDNA library constructed from human sweat glands. Comparing their DNA fragment with a sweat gland library created in the lab of a collaborator, John Riordan, the elusive puzzle piece was found. It was a minuscule, 113-nucleotide fragment contained within the half-million-nucleotide DNA chunk originally isolated by Tsui. The fragment turned out to be a portion of the *CFTR* gene. Using this fragment as bait, the entire *CFTR* gene was ultimately isolated, along with the *CFTR* variant from a patient with cystic fibrosis containing what turned out to be the most common mutation found in the disease. Three papers describing the discovery of *CFTR* were published in the September 1, 1989 issue of the journal *Science*. Williamson graciously congratulated the victors. But accepting defeat must have been difficult for him and his team. The Williamson lab had come agonizingly close. *IRP*, it turns out, is only a hundred thousand nucleotides away from *CFTR*, genomic next-door neighbors. They had jumped off the four-year-long ride to the cystic fibrosis gene a single stop too early.

The most common mutation responsible for cystic fibrosis, the one identified by Tsui and Collins in their initial reports, is a deletion of three nucleotides at codon 508 that leads to the loss of a phenylalanine amino acid residue in CFTR protein. The scientific name of this mutation is ΔF508; Δ is the Greek letter delta, genetics shorthand for a deletion or loss of genetic material. The problem with the mutant protein is that it is processed abnormally in cells. After proteins are synthesized, they fold or assemble into their final three-dimensional configuration through chemical bonds that form between the different amino acids. Protein assembly does not always occur reliably on its own. Like an unwieldy ocean liner needing tugboat support to negotiate a narrow harbor, complex proteins sometimes need guidance to fold into their proper configuration, utilizing other proteins called chaperones. All of this takes place in the endoplasmic reticulum (the protein-processing organelle). Proteins that fail to fold properly with their chaperone aids are targeted for destruction by another protein called ubiquitin. Ubiquitin-bound proteins are degraded and destroyed by cellular enzymes in structures called proteosomes. It is survival of the fittest, at the protein level. The only proteins to survive are the hardy ones–those that fold quickly into the proper configuration, which permits each protein to migrate to its appropriate cellular destination and perform its programmed biochemical function. Ineffectual proteins are destroyed. Abnormally folded proteins that overwhelm the ubiquitin pathway and accumulate in cells are not only a burden, they are downright dangerous. Alzheimer's disease and "mad cow disease" (bovine spongiform encephalopathy) are caused by the toxic effect produced by abnormally processed brain proteins on neurons.

For ΔF508, the loss of a phenylalanine amino acid alters its capacity to fold efficiently, which activates ubiquitin, and the mutant protein is destroyed before it can reach its location on the membrane. It maintains its function as a chloride channel when studied in the laboratory in the absence of ubiquitin, but its misfolding and subsequent inability to migrate to the cell membrane renders it useless in real life. Since its underlying function as an ion transporter is preserved, researchers are looking for drugs that will inhibit the processing defect and allow some ΔF508 protein molecules to find their way to the cell membrane, where they would have the capacity to function properly. One drug that has been found to be effective in animal models of cystic fibrosis is curcumin, an active ingredient of tumeric. Curcumin inhibits chaperones. By blocking chaperone activity, ΔF508 escapes degradation, allowing it to migrate to the surface. Whether this type of treatment will work in humans without having a detrimental effect on the normal chaperone pathway remains to be seen. But the finding illustrates the potential treatments that arise from a deep understanding of molecular and genetic disease pathways.

The ΔF508 variant accounts for about 70 percent of the *CFTR* mutations that can cause cystic fibrosis. Since its discovery, nearly a thousand others have been found in patients, the largest number of mutations uncovered in people with an inherited disorder, testimony perhaps to the survival advantage against severe intestinal infections afforded to carriers throughout human evolution. Of these, only about a dozen variants cause most of the cystic fibrosis in the world. Some are so rare that they have only been found in single families. The common denominator for all

CFTR variants is a reduction in functional CFTR protein expressed on the surface of cells. For ΔF508 and the other most common *CFTR* mutations, no CFTR protein migrates to the cell membrane. For some variants, though, CFTR function is reduced, but not completely abolished, resulting in less severe disease, providing a minority of cystic fibrosis patients with enough exercise reserve to function normally, even allowing some to jog a marathon. This suggests that even if CFTR function were only partially restored in patients with severe cystic fibrosis, it would have a dramatic effect on their quality of life and longevity. One type of debilitating *CFTR* gene mutation, responsible for about 2 to 5 percent of all cystic fibrosis cases, results in the complete absence of CFTR protein production; this severe version of the condition is caused by genetic coding defects called "nonsense mutations." In one of the more remarkable demonstrations of the law of unintended consequences, the aminoglycoside class of antibiotics, which is used to treat the *Pseudomonas aeroginosa* infections that plague cystic fibrosis patients, has been found to partially correct *CFTR* nonsense mutations, previously considered to be one of the three most damaging types of gene-debilitating defects. Whether or not this proves to be useful in treating cystic fibrosis remains to be seen. Kidney damage, loss of hearing, and selection for resistant bacteria would be anticipated problems for patients undergoing long-term aminoglycoside use. But, it is a nice "proof of principle" finding that may prove to be clinically useful at some point in some patients. It also illustrates some of the tantalizing research possibilities that have emerged as a result of isolating the *CFTR* gene. One avenue of research made possible by the discovery of *CFTR* is gene therapy to treat cystic fibrosis victims (see chapter 16).

With the discovery of *CFTR*, it became possible to analyze DNA obtained by amniocentesis or chorionic villus sampling to determine if a fetus is carrying two abnormal *CFTR* alleles, indicating that cystic fibrosis will occur. Prenatal diagnosis for this condition is performed in the same manner as that for sickle-cell anemia, although cystic fibrosis testing poses more substantial diagnostic challenges. While geneticists have only a single mutation to consider in sickle-cell anemia (the beta globin gene codon 6 error), more than a thousand different mutations in *CFTR* are capable of causing cystic fibrosis. However, the ΔF508 variant is found in about 70 percent of cystic fibrosis of carriers, and thus, by testing for that and the other dozen most common *CFTR* variants, it is possible to diagnose cystic fibrosis fairly easily in a fetus with great accuracy in about 95 percent of cases.

The American College of Obstetrics and Gynecology and the American College of Medical Genetics have both recommended that there be widespread screening for the two dozen most common *CFTR* mutations in all couples trying to achieve pregnancy. It is not feasible to screen for each of the thousand rare variants that can cause cystic fibrosis, but exceptions are made for a couple at risk who have already had one affected child where the common variants are not the cause. Even with these limitations in place, though, the logistics of mass screening for cystic fibrosis carriers is daunting. In the United States, more than 4 million babies are born every year. Testing each prospective parent for carrier status is expensive, although it is still less expensive than caring for the hundreds of cystic fibrosis children that would have been born every year if prenatal screening were not carried out. Despite the evidence in favor of universal cystic fibrosis screening in Caucasians, it is not yet consistently practiced in the United States; some centers routinely test, others do not. Centers that do not provide testing or referral services defy the advice of two expert boards, and set themselves up for lawsuits. There have already been several malpractice suits filed by parents who have had children with cystic fibrosis without having been offered the option of prenatal screening.

Prenatal cystic fibrosis screening is not always straightforward. Determining a carrier's status can be complicated. One *CFTR* variant known as R117H does not cause cystic fibrosis unless it is accompanied by a second mutation on the same *CFTR* gene known as 5T. In order for R117H to cause cystic fibrosis, both *CFTR* genes have to contain the double mutation (R117H and 5T), or else one copy of the double mutation has to combine with a classic cystic fibrosis-causing mutation on the other *CFTR* allele, such as ΔF508. If you find this confusing, you are not the only one. Some physicians and genetic counselors have had problems interpreting this prenatal screening genetic data, and several fetuses with R117H but without 5T have been aborted unnecessarily. Despite these and other problems with prenatal screening for cystic fibrosis, hundreds of healthy babies who would not otherwise have been conceived are born every year in the United States, instead of their doomed siblings.

The discovery of *CFTR* has led to an unexpected explanation for the unique persistence of *Pseudomonas aeruginosa* in the airways of people with cystic fibrosis. New ideas about the relationship between them and their most important bacterial nemesis have emerged from an analysis of a mouse model of cystic fibrosis. The mouse model was created by first genetically deleting or "knocking out" one copy of the mouse *CFTR* gene from developing embryos. Mice born with a single *CFTR* deletion were mated to each other to create offspring who do not have any functional *CFTR* genes. In an ironic twist of genetic fate, mice with "engineered" cystic fibrosis actually fare better than humans with the disease, primarily because *P. aeruginosa* does not naturally infect rodents. In fact, to test the effects of this microbe in mice, large numbers of *P. aeruginosa* have to be shoved up their nostrils or introduced at high concentrations in their drinking water. These types of studies show that it is very difficult to artificially induce such infections in genetically conventional mice, but it is easier to do so in mice subjected to *CFTR* "knockout." The upshot of the mouse studies, as proposed by Gerald Pier, is that functional CFTR protein is needed to protect the airways from *P. aeruginosa*.

How does this occur? As described in previous chapters, an initial event in the life cycle of microbes that invade cells, such as viruses, *Salmonella typhi*, and *Plasmodium*, is their attachment to specific molecules located on host membranes. This initiates membrane fusion and internalization of the invading microbe. One important fact about this process that was unknown before the *CFTR* gene was isolated is that some bacterial strains use CFTR protein for cellular attachment–*Salmonella typhi,* for example, the causative agent of typhoid fever. Ingested *S. typhi* are internalized by cells located in the wall of the small intestines, using CFTR as a hook. Indeed, the typhoid bacillus has gone to great lengths to acquire the ability to recruit CFTR protein into its invasion scheme (see chapter 9). Heterozygotes for mutant *CFTR*–people with a single normal copy and a single mutant copy–have less severe typhoid fever because fewer *S. typhi* are internalized, providing the infected gut with a better chance of fighting off the invader.

However, membrane proteins such as CFTR can also be used as part of the host's defense strategy against certain microbes. The interaction between cells lining the respiratory airway and *Pseudomonas aeruginosa* is an example. It turns out that airway cells also use CFTR protein to trap *P. aeruginosa*. But, unlike *S. typhi*, *P. aeruginosa* prefers to grow outside of cells, a favorite site being the mucus secretions of the respiratory airways. Although mucus has antibacterial properties, *P. aeruginosa* is resistant to its inhibitory effect. Epithelial cells lining the respiratory airway use CFTR to remove *P. aeruginosa* from its natural infectious niche; internalization into respiratory epithelial cells is an unfavorable growth condition for this particular microbe. Microbes that are genetically programmed to grow inside of cells often damage them in some manner: for example, HIV kills the CD4 subset of T-lymphocytes, which serve as its host cells; *S. typhi* forces phagocytes to commit suicide; *Plasmodium* induces red blood cells to burst; and so on. Not so for *P. aeruginosa*. This microbe is harmless when trapped inside respiratory epithelial cells. These cells naturally slough off periodically and are replaced by new ones. Worn-out respiratory epithelial cells are propelled from the body, along with the trapped microbe, by coughing.

Consequently, in addition to its role in chloride transport, CFTR protein is a "molecular mop," selectively removing a pesky microbe from the airways, working with respiratory airway cells and the cough reflex to protect us from developing chronic *P. aeruginosa* infections. CFTR has, in a sense, been recruited by the human body as an unconventional component of the innate immune system. In the absence of functional CFTR, *P. aeruginosa* remains in the mucus secretions of the airways, where it grows freely, initiating the chain of events that ultimately leads to lung destruction and pulmonary failure in patients with cystic fibrosis.

Medical discoveries have doubled the life span of cystic fibrosis patients. Isolation of the *CFTR* gene and its many dysfunctional variants has provided family-planning options for couples at risk for conceiving a child with cystic fibrosis, and has encouraged hope for the possibility of a cure by either gene therapy or pharmacological manipulation sometime in the future. But unexpectedly, we have also learned something about human evolution from the genetic discoveries regarding cystic fibrosis. Evolution has provided us with an elite corps of defenders: dedicated phagocytic cells that consume and destroy microorganisms, the complement pathway and antibacterial peptides that limit bacterial growth, and an army of B- and T-lymphocytes capable of recognizing a billion different antigens. The orchestrated antimicrobial attack by these various arms of the immune system is awe-inspiring. However, our genome has also found a way to recruit unlikely soldiers–such as CFTR chloride channels in cells lining the respiratory airway–to serve us in the struggle for survival against infectious diseases. What we have learned from studying people with cystic fibrosis who have lost the ability to utilize the antimicrobial effect of CFTR is that without a few inventive "low-tech" tricks in our genetic repertoire to support the elite immune system against pernicious microorganisms, and perhaps a bit of luck as well, humans and our hominid ancestors might not have survived our millions of years of passage through the microbial world.

CHAPTER 11

The Super Mutation

The relationship between infectious pathogens and protective mutations in the human genome is not always based on a one-to-one connection. A single genetic variant can sometimes influence the body's response to several different microbes. Take for instance the principal cystic fibrosis mutation, $\Delta F508$, which influences the invasion tactics of three bacterial species, *Salmonella typhi*, *Pseudomonas aeroginosa*, and *Vibrio cholerae* (see chapters 8, 9 and 10). In this chapter and the two that follow it, another genetic variant called CCR5-Δ32 (pronounced *CCR5-delta 32*) is considered in the context of its possible effect on three of the world's most destructive pathogenic organisms and the infectious diseases they cause: bubonic plague, smallpox, and AIDS. Although the scientific evidence for a connection between *CCR5-Δ32* and AIDS is the only one of the three that is based on a very firm scientific footing, the chance that there might also be a connection to plague and smallpox is too tantalizing to ignore. Besides, there are historic and scientific lessons illustrated by the accounts of plague and smallpox (which are caused by organisms that are both high on the bioterrorism alert list) that should more than make up for the speculative nature of the links between *CCR5-Δ32* and those diseases.

The fundamental vulnerability of people infected with HIV is caused by the virus's destruction of CD4 cells, a subset of the T-lymphocyte family. CD4 lymphocytes are also called "helper cells" because they help other components of the immune system. Not surprisingly, the loss of CD4 cells dramatically increases a person's susceptibility to all sorts of bacteria, fungi, and viruses. The deterioration of the immune system caused by HIV is usually slow but relentless. With every passing year, about 10 to 20 percent of CD4 cells in a person with HIV/AIDS are lost. Most people have about 1,000 of these immune cells in every thousandth of a milliliter of blood. Most HIV-infected people stay healthy until their counts drop below 200, the immune system's "Mason-Dixon line." Although secondary infections caused by a variety of common and exotic microbes can occur in people with higher CD4 counts, the frequency increases dramatically once the 200 level is breached. The most serious infections found in people with AIDS are: pneumonia from *Pneumocystis carinii*, a common fungus that usually causes a benign pulmonary infection in children; tuberculosis; a chronic infection with *Mycobacterium avian intracellulare,* a cousin of *Mycobacterium tuberculosis*; *Toxoplasmosis gondi*, a parasitic protozoan transmitted by cat feces, which causes brain abscesses; and *Cytomegalovirus,* which can cause blindness and affect the gastrointestinal tract, and liver. The clinical diagnosis of AIDS is defined by the development of any of these bacterial and fungal infections–which are known as opportunistic infections–caused by declining CD4 counts. Much of the improved survival of patients with AIDS can be attributed to the increased proficiency in recognizing and treating these opportunistic infections, and not simply the ability to suppress HIV growth per se.

Not surprisingly, the amount of HIV in the blood (the viral load) is usually proportional to the rate of CD4 cell decline.

Not everyone infected with HIV follows the same pattern of disease progression. Viral loads and the rate of CD4 cell decline differ from one person to another. Some individuals who were infected at the beginning of the pandemic, in the early 1980s, died within a few years; and some are still alive today. A small minority has even been able to sustain their immune system for years without medication. Like most other situations in medicine, there are two extremes and everything in between.

On one end of the progression spectrum was a patient the author once treated, a woman in her early 20s, who developed symptoms of AIDS less than two years after getting infected with HIV. As an intravenous heroin-user since the age of 15, she had managed to dodge HIV successfully for several years; she was also promiscuous, having bartered sex for drugs and thus adding to her HIV risk. But after eight years of living dangerously, she gave up anonymous sex, stopped using heroin, and entered a drug treatment program. She had been tested for HIV every year, urged to do so by her mother (with whom my patient lived), who was vigilant about the disease and insisted that her daughter get tested frequently. She had a string of three or four negative tests before coming under my care. The situation changed when she found a new boyfriend, a man who was about twice her age. In her infatuation with the man, my patient ignored the HIV precautions preached by our counseling service. It turned out that he too had been an intravenous drug user in the past and had a reputation as a womanizer. When it came time for her next HIV test, she was anxious and only agreed to get it done because her mother insisted. This time, the test result was not good. She took the news, not unexpectedly, quite badly, and screamed that she was going to stab "the bastard" in the heart while he slept; it took her an hour to calm down. She was reassured her that it would be years before the immune system would begin to register the effects of HIV, and that there were potent medicines available to fight the virus once treatment became necessary. She could not bear to tell her mother the bad news, so she passed the unpleasant task onto me. Her mother bravely accepted the results of the HIV test.

Deciding when to begin treatment for HIV is part of the "art" of medicine. Treatment is usually delayed until the CD4 count drops below 350 or until an AIDS-related opportunistic infection occurs, regardless of the CD4 count. Doctors can't begin treatment too early, because the antiretroviral medications needed to treat HIV are so difficult to take. There are only two or three good opportunities for keeping HIV under control with different triple combinations of antiretroviral medications before resistant strains begin to emerge. We call the treatment regimen HAART (highly active antiretroviral therapy). Physicians try to avoid wasting a HAART regimen on patients whose CD4 counts are so high that there is little chance that an opportunistic infection will emerge, even without treatment. However, if HAART is started too late, the patient's response is less than optimal. Some patients want to begin treatment as soon as possible, whereas others need more time to understand the illness and the commitment it takes to keep HIV in check. It is especially difficult for some patients who have no symptoms but whose CD4 counts are in the danger zone. Every patient has unique features that have to be taken into account; this is the "art" of the decision-making process in medicine. Once the decision is made to start treatment, the HAART combination of medications has to be taken religiously; otherwise, the virus can bounce back as a morphed, drug-resistant mutant, making the infection more difficult to treat.

There were a few disturbing signs in this patient that made me think that she would need to be started on medication sooner rather than later. At the very beginning of her HIV infection, the baseline CD4 cell count was about 500, which is at the low end of the normal range, and the viral load was very high. When the CD4 count was measured a few months later, it had plunged to the low 400s; and after a year, it was in the 300s. Her illness was progressing rapidly, and I was anxious to start HAART. She resisted at first, since she felt healthy and strong, and was not emotionally ready to handle the burden of the HAART cocktail. But after a few more months, the CD4 cell count had dropped farther, and she experienced her first bout with an AIDS-related opportunistic infection: a bad case of oral thrush, a fungal infection of the mouth that can spread into the esophagus. She finally agreed to begin treatment. Within a few weeks, the viral load dropped and her CD4 count began to rise.

Because she had been tested for HIV every year, we were able to pin down within a fairly narrow time frame the period when she was infected with the virus: it was sometime during the eight or nine months between her last negative HIV test and meeting her boyfriend, who it turned out had known he was HIV-positive. She had been infected with HIV and progressed to clinical AIDS in just under two years.

At the same time, the author was following another patient, a former intravenous drug user who had been HIV-positive for about 10 years. Looking at him, you could hardly tell there was a problem. After a decade with the virus, his CD4 count had remained at about 900, and unless an ultra-sensitive test was done, HIV could barely be detected in his blood. Without any treatment, the virus had not progressed one bit in 10 years. Aptly, but without much imagination, we refer to these HIV-infected patients as non-progressors or slow progressors. These lucky individuals represent a small but measurable minority of HIV-infected people who have managed to maintain their CD4 T-lymphocyte levels in the normal range for many years without treatment.

This patient believed that he was naturally inclined to be resistant to the effects of drugs and disease. He used to brag about the large amount of heroin he shot up every day, all the while managing to hold a job and keeping his family life intact, both rare occurrences for heroin addicts. By the time he came under my care, he had stopped using heroin, was taking vitamins and eating properly, exercised regularly, and had reestablished a connection to Catholicism–life-style changes that, together with his "natural resistance" to all drugs and diseases, he felt were responsible for his extraordinary capacity to keep his HIV infection in check. These were all good health choices, no doubt; but in his case, luck played the key role. He was the fortunate recipient of the genetic mutation CCR5-Δ32 that has the remarkable capacity to keep HIV's destructive effect on the immune system in check.

As described in previous chapters, viruses and other intracellular parasites usually need to hook onto a receptor on the cell's surface in order to gain entry. For HIV, the cellular hook is a cluster of receptor molecules called CD4, CCR5, and CXCR4, to which the virus binds, initiating fusion with the cell membrane and viral entry. The receptor cluster is found on the HIV target cells–the CD4 subset of T-lymphocytes and macrophages. The CD4, CCR5 and CXCR4 receptors are normally used by immune cells to snatch chemokines (protein factors released at sites of infection and inflammation). Their purpose is to attract lymphocytes and phagocytes to areas of the body in need of immune defenders. The cascade of effects that chemokines have on CD4 T-lymphocytes is initiated through their initial brief interaction with CD4, CCR5, and CXCR4 receptors, which triggers an intracellular signal transduction cascade culminating in immune cell activation. These chemokine receptors have been usurped by HIV, and are so essential to the virus's invasion plan that researchers and pharmaceutical companies are trying to exploit some way of interfering with its ability to latch onto them, as a potential AIDS treatment. One strategy is to bombard infected patients with receptor decoys, analogs of the receptors themselves, to divert HIV's attention from its natural target on CD4 cells and macrophages. Another is to place a barrier between HIV and its target cells by blocking elements of the receptor complex. These novel ideas are compatible with cues that have been provided by nature.

Approximately 20 percent of European Caucasians harbor a mutation in the CCR5 gene that reduces HIV's ability to latch onto CD4 cells: the CCR5-Δ32 variant. People with CCR5-Δ32 have a 32-base deletion of genetic information from the coding region of the affected *CCR5* gene, rendering the encoded receptor protein completely useless. The variant gene is not found in East Asians, Native Americans, or African blacks, unless there has been some Caucasian admixture. The absence in Africans suggests that the CCR5-Δ32 mutation occurred after the migration out of Africa more than a hundred thousand years ago that resulted in the emergence of distinct racial groups. Although it would seem that the loss of a chemokine receptor would adversely affect inflammatory and immune responses, inheriting CCR5-Δ32 has no known physiological consequences. One reason may be a built-in redundancy in the chemokine pathway–that is, if one component of the system fails, others may be able to compensate. In fact, in the 21st century, there is a distinct advantage to inheriting the CCR5-Δ32 variant: one copy delays the onset of AIDS in people infected with HIV by about two or three years. Inheriting two copies, which occurs in about 1 percent of Caucasians, provides nearly complete resistance to the deadly virus. By reducing the number of binding sites that HIV needs to gain cellular entry, progression to AIDS can be abated.

Analysis of CCR5-Δ32 shows very striking differences in frequency among different Caucasian populations, with a distinct North-South distribution. The highest frequency is found in Northern Europeans, such as in Sweden, where about 28 percent of the population carry a single copy. The lowest frequency is in Southern Europeans, in the Mediterranean region, where it's found in approximately 5 to 10 percent of the population. The initial genetic analysis of the mutation revealed that it may have originated about 700 years ago, although such determinations are difficult to make precisely with any great accuracy. According to Gerard Lucotte, a French molecular geneticist, the North-South European distribution and extensive genetic analysis suggest that CCR5-Δ32 originated in a single progenitor in Northern Europe, and spread south by invasion and intermarriage during the Viking era, from the 9th to the 11th century. This individual unknowingly carried a mutation that may have saved the lives of millions of descendants.

There are several explanations that could account for the stark difference in allele frequency across different ethnic lines. One is genetic drift, a phenomenon that is the exact opposite of Darwinian selection. Genetic drift is an increase in the frequency of a genetic variant caused entirely by chance. This can occur if a catastrophic event unrelated to genetic fitness befalls a small inbred population, and among the survivors is someone who happens to have a particular genetic variant. Say, for example, that in a village of 500 people, a devastating earthquake kills all but 50, one of whom carries a unique genetic variant. From an initial frequency of 0.2 percent (1 in 500 individuals), the unique variant is suddenly found in 2 percent of the population after the catastrophe, a 10-fold increase. Assuming that the variant is physiologically neutral and did not affect survival in earthquakes, non-selective expansion has occurred. It is estimated that 99 percent of the millions of genetic variants present in the human genome expanded in this fashion. Most are non-functional mutations occurring in introns or intergenic DNA.

However, the fact that the CCR5-Δ32 mutation causes a functional change in a protein that influences viral fusion and affects immune responses suggests that a Darwinian selection process caused its great expansion in Europeans. The mutation could not have expanded by survival of the fittest caused by HIV, because the virus originated in Africa and did not emerge as a human pathogen until the late twentieth century. European history and the apparent age of CCR5-Δ32 point instead to two cataclysmic events that could have selected for the cytokine receptor gene variant: the pandemics of bubonic plague and smallpox.

CHAPTER 12

Natural-born Killer

Dead rats by the thousands, littering homes, markets, and streets, signaled the beginning of a bubonic plague epidemic. Hordes of the black rat (*Rattus rattus*)–the predominant rodent species in medieval and Renaissance Europe–became sickened with the disease and would abandon their usual cautious behavior and stagger out of their dark enclaves into broad daylight. Instead of the comfort and safety provided by scurrying along the sides of walls, they would weave about in the open in full view, sickly-looking creatures with glazed eyes and disheveled fur. Sometimes, while moving across the roofs of buildings, they fell dead onto the streets below. Other small mammals–mice, moles, foxes, even dogs and cats–could also be infected and succumb to the same fate.

In the three known widespread pandemics that occurred in a 550-year time period between 1348 to 1894, and in numerous local epidemics in between, bubonic plague killed about 200 million people, mainly in Europe and Asia. It was called "the pestilence" by the people of medieval Europe, and along with the disease they called the "small pocks" (smallpox) and cholera, it shared the dubious top spot for the most feared of all infectious diseases.

Symptoms of bubonic plague begin, like so many other infectious diseases, with flu-like features: fever, weakness, chills, and headache. Sometimes, a few tiny red spots appear on the arms, legs, and face that form crusting lesions after a couple of days. This was a clue to the mode of transmission of plague, which would not be understood until the beginning of the twentieth century. But the symptom that marked the disease as something quite different from other infectious afflictions was the eruption of grotesquely enlarged painful lymph nodes, known as buboes. These swellings usually appeared in the groin, but sometimes showed up in lymph nodes under the armpit, in the neck, and behind the ears. The term "bubo" is derived from the Greek word for groin, *bubon*. Buboes could grow to the size of a fist in a single day. They turned black from hemorrhaging, which, along with the skin hemorrhages that appeared in terminal cases, provided bubonic plague with its menacing moniker, the "Black Death." Buboes sometimes swelled so precipitously that they would burst spontaneously from the pressure, like an over-filled water balloon, temporarily relieving the discomfort of having an engorged mass, but causing further damage by releasing infected pus into the surrounding tissue. Bubonic plague was fatal in about 50 percent of all cases. Sometimes the infection disseminated through the blood and traveled to the lungs, causing pneumonia, a form of the disease called pneumonic plague, which was nearly 100 percent fatal. Worse, victims of pneumonic plague could bypass rat carriers and spread the infection directly to others by coughing bacteria-filled phlegm.

Although plague could strike anyone, women who were kept secluded at home had the highest risk of contracting the infection, and pregnant women and their fetuses had the highest mortality. People in port cities were usually the first to get the disease. Death from secondary infections or gangrene of the legs occurred frequently. Survivors were often disabled because their extremities were damaged by lymphatic obstruction, causing swelling and disfigurement resembling elephantiasis.

The first recorded plague epidemic occurred in the middle of the 6th century and spread throughout the East Roman Empire of Byzantium, also known as the Byzantine Empire. It began in 542 at Pelusium, a port city in Egypt, which probably acquired its initial cases of the disease from the interior of Africa through slave and ivory traffickers, or rather, the rats that traveled with them. A year later, it arrived in Byzantium's capital, Constantinople, the easternmost city in continental Europe. According to scholar Timothy Bratton, it crossed the Mediterranean with the Egyptian corn fleet, a buffet feast for rats. The first pandemic was called the "Justinian plague" (named for Justinian I, the Roman emperor at the time). It took the lives of as many as 25 percent of the people in the Eastern Empire and contributed to the decline in power of the Western Empire after the fall of Rome a hundred years earlier. So many lives were lost that there were not enough men to fully man the armies. Parts of the empire fell to Arabs and nomadic tribes who were spared the disease's devastation. Their lives in the desert placed them at a safe distance from the sea traffic that spread the plague from one port city to another. Persia was also weakened by the plague and easily fell under a barrage of attacks by nomadic muslims, who emerged from their desert bases relatively unscathed by the disease. The tax base diminished because of reduced productivity; monasteries were depopulated; and there were not enough people to harvest crops. Ships empty of crews sometimes turned up, the dead apparently having been thrown overboard, while the dying jumped in an attempt to escape the mysterious scourge on board their vessels. In Constantinople, thousands died daily at the peak of the epidemic. During a four-month period, as many as 50,000 to 60,000 people died, out of a population of approximately 300,000. Nearly half the Mediterranean world perished from the formidable triumvirate of bubonic plague, crop failures, and warfare.

Plague remained endemic in the region for another 300 years, although without the burst of infectious intensity seen in the original infestation. As with so many other epidemic infections, there is a drop in the mortality rate as local immunity is generated in survivors and genetic selection for protective mutations emerge. Also, for zoonoses—infectious diseases spread by animal carriers—an epidemic can die out because of a sharp decrease in the carrier population.

Observant onlookers noted that the disease did not usually jump from person to person (except in cases where the lungs were affected), which differentiated it from highly contagious diseases such as smallpox and measles. Victims of plague appeared suddenly, seemingly out of nowhere. Attendants of the sick and dying had no added risk of acquiring plague following their humane intervention. This was an epidemic with none of the characteristics of other infectious diseases, except for the incomprehensibly high mortality rate.

The bubonic plague remained relatively quiescent for centuries and then, between 1348 and 1350, exploded throughout Europe. The Black Death entered Europe via trade routes between Asia and the Black Sea, through the city of Kaffa. The Tartar army besieged Kaffa at the time and suffered many plague casualties. They may have fueled the epidemic when their leader ordered that the dead be catapulted over the walls of the city. Fleeing Genoese traders brought the disease back to Sicily, after which it was quickly spread throughout the rest of the continent by merchants and traders. It was the second disaster to strike Europe in the 14th century. In 1316, crop failures occurred because of a severe global cooling, the so-called little Ice Age. The low nutritional status of Europe's poor may have contributed to the plague's high mortality.

It was a disaster of monumental proportions, the single most destructive biological event in human history; as many as 20 million people out of Europe's population of 60 million died, most within about a thousand days.[19] The population declined so precipitously that it did not reach the same level again until two centuries later. The great cities of Italy were decimated. In France, Pope Clement VI consecrated the Rhone River so that the dead could be thrown in with a proper burial. In England, nearly 50 percent of the clergy died. As recounted by a 14th-century London cleric named William Langland in his spiritual epic poem *Piers Plowman*, "For god is deaf nowadays and will not hear us."[20] The effects of the disease threw Europe into a social, psychological, and economic tailspin that lasted about two centuries, only ending with the emergence of the Renaissance.

---

[19]The Spanish Flu epidemic of 1918 took more lives, but the overall case fatality rate was lower—about 1 percent. HIV has also affected more people than did the Black Death; but in most countries, except for some African nations heavily infected with HIV, the percentage of the population affected is in the single digits.

[20]Quoted in Norman F. Cantor, *In The Wake of the Plague: The Black Death and the World It Made* (New York: Simon and Schuster, 2001).

Medieval physicians trying to understand the causes of the epidemic were stymied. Medicine at the time was still under the influence of the Greek physician Galen. Bloodletting, purgatives, and observing the color of urine to determine the state of health were common practices of the day. The most common explanation was that a miasma was the source of plague. Those who could leave migrated to the countryside. Rich people covered their windows with elaborate, heavy tapestry to help prevent intrusion of the deadly miasma into their homes—a marketing opportunity for artists and weaver guilds. Frequent bathing was viewed as dangerous; it was believed that the practice opened pores to plague and other diseases. Covering the mouth with a scent-soaked handkerchief was thought to help counter the deadly miasma, and so perfume makers flourished. Astrologers blamed the position of the planet Saturn in the house of Jupiter. Clergymen blamed it all on a punishment from God heaped onto a sinful Earth.

Jews were accused of spreading plague by poisoning wells. Under torture, they were forced to "admit" their crimes or accuse others. This triggered hundreds of pogroms. Throughout Europe, Jews were beaten to death, burned alive, or drowned in the very wells they were accused of poisoning. In Strasbourg in 1349, half the Jewish population—about 2,000 people—were burned to death after plague came to the city. In Esslingen, Germany, Jews committed mass suicide by setting their synagogue on fire with themselves in it, rather than submit to their inevitable slaughter at the hands of angry mobs. In Mainz, Germany, 6,000 Jews were burned to death by such mobs. In some towns, Jews were killed as a preventive measure, even before the arrival of plague. Some of the attacks were instigated by people who had incurred large debts to Jewish bankers, using the plague to incite their anti-Semitic neighbors to violence, as a way of erasing their loans. The Pope tried to intervene on behalf of the Jews, to no avail.[21]

---

[21]The fallacy of poisoned wells was sufficiently entrenched in the consciousness of Europeans that its appearance in Philip Marlow's play "The Jew of Malta," written in 1588, hardly caused a stir in Elizabethan England. In the play, Barabas, the Jewish anti-hero cynically declares, "sometimes I go about and poison wells."

It was a grim harbinger of the holocaust that would occur in 20th-century Europe, committed by the very descendants of survivors of the Black Death.

The labor shortage created by the plague provided peasants with increased leverage, which led to improved wages, paving the way for the feudal system's eventual breakdown. The working poor's standard of living actually improved, since the Black Death created a worker's market. Fearing out-of-control costs, rich English landowners were able to get legislation enacted late in the 14th century to help keep wages modest, which led to the peasant revolt of 1381. The labor shortage caused by the Black Death also created a need for maids to help run the households of the Florentine upper class. Young Tartar, Bulgarian and Armenian girls were sold by their families and slave traders as indentured servants to help fill the gap. When the girls matured they were sometimes impregnated by their owners or his friends and relatives.

Aside from the economic opportunities it created for survivors, there was one other benefit to the outbreak of the Black Death: it wiped out the religious orders that housed lepers, effectively eliminating from Europe the historic chronic infectious disease leprosy, which can cause facial disfigurement and deformed limbs.

The destruction of human life by bubonic plague might have had another inadvertent outcome. According to the geologist and author William F, Ruddiman, plague might have had an effect on greenhouse gases. Measurements of atmospheric carbon dioxide and methane in core samples taken from deep ice sheets in Greenland and Antarctica suggest that these gases, which are responsible for global warming, began to rise well before the modern industrial era as a result of the invention of agriculture in the Mesopotamia and China. Domesticated animals emit methane gas from their intestinal tracks. Carbon dioxide levels began to rise with deforestation, which was needed to clear natural forests to create vast farms. The fallen trees were either burned for heat, or left to decay, both of which emit carbon dioxide. There have been several reversals in the tendency for carbon dioxide and methane levels to rise, two of which occurred at around the time of the Black Plague and the Justinian Plague (another coincided with the introduction of smallpox into the Americas which killed most of the indigenous population, as described in the next chapter). According to Ruddiman, massive depopulation caused by these biological catastrophes temporarily reduced global farming activities and slowed down the greenhouse gas emissions.[22]

Plague persisted for three centuries, periodically erupting in small local outbreaks, although with nothing of the devastation that occurred during its initial debut in 1348. Then, another terrible epidemic broke out in 1665. It was called the "Great Plague."

In London alone, the Great Plague claimed a 100,000 lives. The epidemic was initially confined to the poorer sections of the city. It was ignored by officials when there were only scores of dead among the poor. The population began to panic, however, when the number of deaths increased to the hundreds, then thousands a week. The rich and the not-so-rich fled the city. King Charles II abandoned Windsor castle for more rural environs. There were rumors that dogs and cats were spreading the disease, so the Lord Mayor ordered them destroyed. About a quarter million of them were killed, effectively removing the only predators against the true culprit, the black rat. Londoners tried everything to ward off the disease, from sniffing herbs to carrying posies. Tobacco was a popular choice, so much so that even children were encouraged to smoke, sniff, or chew it. The epidemic spread to surrounding towns. Cambridge University had to close down temporarily, forcing its star student, Isaac Newton, to seek refuge back home (during which he invented calculus). In the village of Eyam, plague arrived in a parcel of flea-infested clothing. However, when the first cases of plague occurred there, the town rector convinced people to stay, fearing that they might transmit the pestilence to their neighbors in nearby villages. Their heroic, self-imposed quarantine helped contain the epidemic, but at a very high cost: about 80 percent of the town's citizens died. By the end of the year, 15 percent of Londoners had perished. Only a cold winter and the terrible fire of 1666 that destroyed much of the city and killed tens of thousands of rats finally put an end to the epidemic there.

---

[22]William F. Ruddiman. How did humans first alter human climate. Scientific American March 2005, 46-53.

The science of microbiology invented by Koch and Pasteur eventually solved the mystery of plague. It happened when two adventurous microbiologists traveled to the Far East to investigate the world's third and last major bubonic plague pandemic, which originated in rural China in the early 1890s and rapidly spread east to Hong Kong and south to India via sea and land trade routes. The outbreak would ultimately prove fatal to tens of millions of people.

The first cases in Hong Kong were diagnosed in May 1894 by James A. Lowson, a Scottish physician who was acting superintendent of the Civil Hospital in Hong Kong. Hong Kong's poor lived in deplorable conditions, in small, dirty, sunless, overcrowded homes, ripe for an outbreak of infectious disease. Most dwellings were also home to flea-infested rats. Within weeks of its arrival, the epidemic was out of control. Row after row of Chinese lay dying on mats in the sparse hospitals. Families died together at home, with no survivors left to inform the authorities. The smell of decaying bodies permeated the warm spring air.

Racial hatred on both sides fueled rumors about the outbreak. The British were largely spared the plague's ravages, evidence to them that squalor and what they perceived as the Chinese people's moral decrepitude were responsible. Filthy immigrants from the mainland were blamed for bringing the sickness to the island. Some Chinese blamed the British and claimed that the plague outbreak was an act of a vengeful god striking out against the rumored practice by Peak Railway of using bodies of Chinese babies as supports for the tram line that ran from Victoria Peak to the business district. Racial tension also surfaced in British-controlled India. Health officials quarantined plague-ridden areas and sprayed lime wash in affected tenements, displacing and inconveniencing thousands. Officials banned the yearly pilgrimage of Muslims to Mecca, and Hindu women riding the railroad objected to having their armpits examined for buboes. British anti-plague fighters were attacked and sometimes killed.

Lowson tried to take charge, but his influence with the Chinese population and local authorities was weak, due to his air of British imperial superiority. He was a tall, athletic man, brimming with confidence and arrogance. In fact, some of his ideas were good. When he became aware of the epidemic in China, he wanted to have all vessels coming in from Canton inspected, but was denied permission by the governor. Lowson blamed the inertia of local medical officials for the outbreak in Hong Kong.

As the dead piled up, British soldiers were ordered to clean up the mess. They went on house-to-house searches looking for the sick to take to the hospital in order to isolate them from the general population, and cleaned and disinfected along the way. Within weeks, the hospital was overrun, and warehouses and factories had to serve as temporary housing for the sick and as morgues for the dead. Lowson himself had tried to find the cause of plague but lacked the necessary expertise in the new science of microbiology to make any headway.

Like the search for the cholera bacterium a decade earlier, the technical skills needed to solve the mysterious infectious disease came with the arrival of competing research teams dispatched by Pasteur and Koch. The Koch contingent was headed by a Japanese microbiologist, Shibasaburo Kitasato (1852-1931), who had trained for seven years in Koch's laboratory in Germany. During that time, he developed an international reputation for his discovery of the bacterium responsible for tetanus, *Clostridium tetani*. Kitasato helped develop the low-oxygen conditions necessary for culturing the anaerobic organism in the laboratory. In the course of his work, Kitasato also discovered that the symptoms of tetanus were caused by a toxin produced by *Clostridium* rather than the direct actions of the bacterium itself. In 1890, working in collaboration with Emil von Behring (1854-1917), he found that serum taken from the blood of tetanus victims contained substances that neutralized the action of toxins produced by the tetanus bacillus; these are now known as antitoxin antibodies. He had deservedly earned a reputation as a methodical, careful scientist, and was viewed as one of the best microbiologists in the world, a possible successor to Koch himself.

Pasteur sent Alexandre Yersin (1863-1943), a Swiss citizen who had trained as a physician in Lausanne and Paris. Too retiring to be a practicing doctor, Yersin gravitated toward pathology, a profession that inadvertently led to his association with Pasteur. While dissecting the spinal cord of a rabies victim, Yersin exposed himself to the deadly virus when he cut his finger, an occupational hazard for pathologists in the past and for veterinarians today. He immediately sought Pasteur to receive the rabies inoculations that had just been developed. Yersin was so impressed with the work at the Pasteur Institute that he stayed on for training. There, working with Emil Roux (1853-1933), he made an important discovery on *Corynebacterium diphtheriae,* the bacterium that causes diphtheria. They noted that the bacterium was easily cultured from the thick, choking muck that coats the throat of diphtheria victims, but it could not be found in the heart, which fails in fatal diphtheria cases. They postulated that heart tissue was damaged by a bacterial toxin and not from the direct effects of the infectious microbe itself. They eventually recovered a toxin from the urine of an infected patient.

The diphtheria and botulism toxins were the first bacterial poisons ever discovered. Later, Kitasato, von Behring, and Paul Ehrlich (1854-1915) immunized guinea pigs and other animals with heat-attenuated toxin to produce an anti-diphtheria antitoxin. The experiments provided the scientific framework for the first diphtheria vaccine, a cousin of which is used today as part of the D.P.T. vaccination series given to all infants in the U.S.

But Yersin had a passion for travel and was bored with the social life of late-nineteenth-century Paris. In 1890, to the dismay of his colleagues, Yersin left the Pasteur Institute and the comforts of Parisian life to become ship's doctor on a vessel bound for Indochina (now Vietnam), which ultimately became his adopted home. Yersin spent four years studying tropical diseases and exploring the interior of the country (he was an admirer of African explorer David Livingston). During his explorations, Yersin gained firsthand experience with the diseases of poor countries when he contracted malaria and dysentery. Indochina was also where he saw his first cases of bubonic plague and became aware of the disease's potential magnitude. An epidemic with the impact of the Black Death had the potential to kill a hundred million people if it ever took hold in the crowded cities of the Far East and Southeast Asia. He got the opportunity to study the problem in 1894 when he joined the Colonial Health Service and was dispatched, on Pasteur's recommendation, to investigate the outbreak in Hong Kong.

Both Kitasato and Yersin arrived in Hong Kong in early June. Kitasato came with a team of investigators and assistants; Yersin arrived three days later with a single servant and a microscope. Lowson immediately disliked the shy Yersin, his polar opposite, and favored the more extroverted Kitasato. Lowson showered the Japanese scientist with laboratory facilities, resources, and access to patients. They quickly became friends and colleagues, meeting with each other frequently for private dinners and daily research updates. Yersin was completely ignored. In fact, Lowson undermined Yersin by withholding access to patients and corpses and by denying him laboratory space. Yersin was unable to begin his investigations for another week.

Yersin appears to have met Kitasato only once, while the latter was performing a dissection on a plague victim. To Yersin's surprise, Kitasato was gathering culture material from the blood and internal organs, but did not pay much attention to the buboes, plague's most distinguishing clinical and anatomical feature. Although swollen lymph nodes are found in many infectious diseases, none are as massive as the swelling seen in plague. It was an error in scientific judgment that would haunt Kitasato for the rest of his life.

To show how quickly a scientist could establish a name for himself in the early days of microbiology–when so many of the fundamental discoveries had not yet been made–Kitasato isolated what he believed to be the causative agent of plague within a few days. One of the most infamous microbes in human history lay hiding in plain sight, waiting to be found by anyone with a good microscope and basic skills in bacterial culture techniques. Kitasato isolated his bacterium from the blood of a plague victim, and, with Lowson's help (since Kitasato spoke some German, but very little English), wrote a paper describing his achievement and quickly dispatched it to the journal *Lancet* on July 7. It appeared in the August 25, 1894 issue, only ten weeks after Kitasato's arrival in Hong Kong–lightning speed for making a scientific discovery and getting it to press, by today's standards. It was Kitasato's first and last English-language publication.

Yersin, after the critical one-week delay imposed by Lowson, was finally permitted to build a small hut adjacent to the hospital in which he could set up a makeshift laboratory. However, he was still denied access to patients and had to bribe British sailors for a few clandestine moments in the morgue. Yersin made the best use of his limited time. Convinced that the huge groin swellings held the secret of bubonic plague, he dissected a bubo, removed some fluid and tissue, and returned to his hut, where, looking at his very first clinical sample under the microscope lens, he saw the bacterium that had nearly destroyed medieval Europe. After that, Yersin's research progressed rapidly. He cultivated the bacterium and injected it into experimental animals, where it caused a fatal infection resembling plague. The animals were swarming with the same organism that he had isolated from a human patient's buboes. His experiments satisfied "Koch's postulates," even before Koch's own disciple, Kitasato, was able to do so. Yersin sent slides and cultures back to Paris and quickly dispatched a paper, written in French, that was published in the Pasteur Institute's own journal. Yersin named the new organism *Pasteurella pestis.*

Thus, depending on whether or not English or French was your scientific language of choice, either Kitasato or Yersin was credited with discovering the plague bacillus. However, there were discrepancies in Kitasato's original description of the organism in his *Lancet* paper, compared to Yersin's report, raising suspicions that the Japanese investigator had in fact isolated the wrong bacterium. Differences between the organisms isolated by Kitasato and Yersin were noted in their staining characteristics, their appearance as colonies growing on agar plates, their optimal growth conditions, and whether or not the organism was motile. Soon, Kitasato's own Japanese colleagues were questioning his initial findings. When Kitasato saw Yersin's results using buboes as a source of infected material, he switched from his focus on blood and realized that he had made an error. The plague bacillus is most concentrated in the giant buboes and not in the blood. The first report sent off by Kitasato had been a description of a contaminant, possibly *Streptococcus pneumoniae,* the most common cause of acute pneumonia, which emerged as a secondary infection in some terminal plague cases. Ironically, it may have been Yersin's primitive working conditions that helped him isolate the causative organism first. Kitasato's lab was equipped with incubators in which he could grow bacteria under standard conditions, which usually include a temperature set at 37 degrees centigrade, the average temperature of the human body and the optimal growth temperature for most human pathogens. Yersin was not provided with an incubator. He could only grow his cultures under ambient conditions, between 5 and 10 degrees centigrade cooler than body temperature. However, unlike most human pathogens, the plague bacillus grows better in the lower temperature range. By using the standard-type incubator that microbiologists customarily employed for bacterial cultures, Kitasato inadvertently promoted the growth of a contaminant at the expense of his real target. Although privately, Kitasato acknowledged the possibility of error, and had switched to an analysis of buboes for all his subsequent work, he avoided any public declaration of the mistake. His work on plague and the publication of his paper in *Lancet* had added greatly to his fame, and he did not want to lose face. Lowson stubbornly maintained to the day of his death in 1935 that Kitasato, with his help and support, was the true discoverer of the plague bacillus.

Yersin, celebrated as the co-discoverer of the *Pasteurella pestis*, ignored the controversy that followed. He soon returned to Indochina, where he continued to work on plague and tropical diseases, and pursued his hobbies of astronomy and agronomy. The shy Yersin never married, and he shunned scientific meetings, which helped delay the true history of the plague bacillus discovery for decades. His research in Indochina was financed by profits gained from his coffee- and corn-growing enterprises. He also helped bring rubber trees to the region and is credited with introducing cinchona to the malaria-infested country so that quinine could be cultivated locally.

Yersin died in Indochina in 1943 but he almost did not make it back to his adopted country for the final three years of his life. His last visit to France in 1940, where he attended a meeting at the Pasteur Institute, was nearly a permanent one. Yersin barely escaped Paris in one of the last planes to leave the country after the Germans began their occupation of France. Six hours after his departure, the invading Nazis closed the airport.[23]

---

[23]The close call between Yersin and the Nazis was not the only plague/World War II connection. Around the time of Yersin's departure from France, invading Japanese forces intentionally released plague infected fleas from airplanes over the Chinese city of Ningpo. About 500 people were killed. Other cities also suffered from this act of bioterrorism. The records were destroyed, so the exact toll is not known.

Eventually, Yersin was recognized as the true discoverer of the cause of bubonic plague. His name became permanently associated with the disease when the causative organism was renamed *Yersinia pestis* in 1970. Yersin's tomb is still honored in Vietnam, and, despite the removal of most vestiges from the French colonial era, many streets and buildings named for the unassuming Swiss physician-scientist continue to carry his name.

It was left to another Pasteur disciple, the Frenchman Paul-Louis Simond, to complete the picture of how bubonic plague was transmitted to humans. An association between plague and rats had been recognized for years. The Chinese had observed that contact with dying rats seemed to transmit plague, and that people ran from their homes during the epidemic if dead rodents were found. In a Bombay wool factory, laborers ordered to clean up dead rats came down with plague, while other workers did not. Recognizing the association, Chinese officials even offered a bounty for rat tails, but the program failed because people would cheat by importing rat tails from the mainland in order to obtain the bounty money. After isolating the plague bacillus from humans, Yersin had analyzed some of the dead rats strewn about the streets of Hong Kong during the epidemic, and had found the same organism in their bodies. However, the mode of transmission from rats to humans was not clear. Airborne transmission, rat-contaminated food, or entry from dirt through skin abrasions were some of the possibilities considered.

While investigating the rat connection in 1897, Simond made an important observation. He noticed that exposure of humans to the corpses of rats that had been dead for many hours, whose bodies had cooled, did not transmit the sickness. This gave Simond the idea that fleas might be the vector that carried bubonic plague from rats to humans. Fleas feed off the blood of mammalian hosts and are attracted to warmth. Once an animal dies and its body begins to cool, fleas jump to another warm host, using their sensitive heat detectors as a guide. The rat flea, *Xenopsylla cheopis,* prefers the small, low-to-the-ground, hairy bodies of rodents. But, with the sharp reduction in the rat population caused by plague, it will jump onto the nearest warm-blooded mammals, which for urban rats are humans and their pets.

To prove that rat fleas carry and transmit the plague bacillus, Simond performed a series of simple experiments, which carried some personal risk. Using a long forceps, he grabbed a recently deceased rat by the tail and dropped it into a bag of soapy water to kill any fleas on the rat's body. He then dissected the insects under a microscope, in the same manner as Robert Ross (who found *Plasmodium* swarming inside the *Anopheles* mosquito's intestinal tract and salivary gland; see chapter 3). The flea's intestines were infested with the same bacillus that Yersin had discovered. Simond then put a dying rat at the bottom of a tall cage, into which he suspended another cage housing a healthy rat. As long as the healthy rat was suspended above the dying rat with four inches of space in between (the maximum height that rat fleas can jump), the suspended animal was safe; but when the space was less than four inches, and thus within the reach of hungry fleas, plague was transmitted. Moreover, plague-infected rats that had had their fleas carefully removed did not transmit the disease to other rats under the same conditions. However, fleas taken from dying rats were able, on their own, to transmit the plague when they were put into a cage with a healthy animal. Simond had clearly established the mode of transmission, although his findings were dismissed for years.

The role of fleas explained the tiny red specks that some people had observed on plague victims through the ages: these were flea bites. Flea transmission also explained the pattern of bubo formation. They are usually found in the groin because the low-lying rat fleas would jump onto a person's leg, where the fleas would bite, sending the deadly microbe to the nearest lymph-node chain, which is located in the groin. Buboes in the armpit, in the neck, and next to the ear occurred after fleas jumped from dying rats onto the arms and faces of people who slept on the floor and bit them in those areas of the body.

Simond advised people to sleep on beds and not on floor mats. He also urged health officials to prevent rats from leaving boats, which he now realized was the primary mode of transmission in spreading plague along water routes. These simple ideas ran counter to public-health measures at the time, which focused on cleaning with bacteria-killing disinfectants, according to the principles of hygiene developed by Joseph Lister in the 1860s. Although Lister's disinfection methods revolutionized the practice of surgery and obstetrics, it had no effect on a bacterium carried by a jumping arthropod. The connection between fleas and plague transmission was only fully accepted a few years after Simon's death in 1947.

Human-borne fleas, *Pulex irritans,* are also able to transmit plague from human to human.

Plague finally came to in the Unites States in 1899 when a ship from Hong Kong with two plague victims and infected rats on board docked in San Francisco. A small epidemic subsequently broke out in Chinatown. The Board of Health tried to quarantine the affected section of the city, but many businessmen and the Chinese community objected. California's governor refused to believe that an epidemic existed. Despite the public denial, thousands of rooms were cleaned by local officials in an effort to reduce the rat population. A citywide extermination effort, including the offer of a bounty, resulted in the killing of hundreds of thousands of rats. The small outbreak was beginning to wane when it was given new energy in the 1906 earthquake that destroyed much of the city. The collapse of buildings, new housing construction, and the creation of makeshift refugee camps led to an explosion in the rat population, which caused a spike in the number of plague cases. Within a couple of years, the outbreak was over, but not before scores of people died. After the plague bacillus arrived in the United States, it jumped–literally and figuratively–from rats to indigenous rural mammals, via fleas. Prairie dogs in the Southwest are now the prime carriers of *Yersinia pestis* in the United States. Every year, several people die of plague in the Southwest after coming in contact with infected animals. It can be treated with a number of different antibiotics, but it has to be recognized first. The average physician examining a patient who has a serious acute infection does not immediately think of plague, especially the rare pneumonic variety, as a possible diagnosis.

Other than San Francisco, bubonic plague never gained a foothold in the larger urban areas of the United States, where rat overpopulation could have caused a nightmare of an epidemic. New York City was spared by a combination of good luck and vigilant inspection. Despite being free from a major plague epidemic, city dwellers in the United States who see filthy rats scurrying about scavenging for food are filled with revulsion, perhaps associating the creatures with their role in the near-destruction of European civilization.

There are about 1,000 plague fatalities a year throughout the world. One of the most plague-ridden areas is the central Asian republic of Turkmenistan, where the disease is carried by a gerbil, *Rhombomys opimus*. During the Soviet era, an outbreak there killed several hundred people-a single index case, a nomad hunter, was bitten by infected fleas and he subsequently spread the disease to his village by the pneumonic route. The situation is made worse because of Turkmenistan health policies: the dictator Saparmurad A. Niyazov has outlawed the diagnosis of plague and all other infectious diseases, including mentioning them by name.[24]

Plague can still strike fear in our hearts. Because of the threat of bioterrorism, the close scrutiny of stocks of *Yersinia pestis* maintained by plague researchers has been magnified by a heightened sense of concern, even paranoia. *Y. pestis* can only be handled in specially contained facilities located at a few medical and research centers. Recently, a well-known plague researcher, Dr. Thomas Butler, the chairman of microbiology at Texas Tech, was indicted and tried for mishandling thirty vials of the plague bacillus, which he brought surreptitiously into the United States from Tanzania for research purposes. Butler, whose actions triggered a bioterror scare, was found guilty of illegally trafficking the dangerous microbes and then trying to cover up his possession of them, for which he was sentenced to a hefty fine and a two-year jail sentence (the judge could have imposed a stiffer sentence but reduced it after considering Butler's service to humanity).

---

[24]From An illegal outbreak of plague-by Wendy Orent-LA Times, 2004).

The arrest of Dr. Butler is a far cry from the more cavalier attitude toward the plague bacillus in the past. The days of shipping the bacillus through the mail, as Yersin did when he sent samples from Hong Kong to Paris, are over. Ludwig Gross, a Lasker award-winning virologist who died in 1995, wrote about a mishap involving *Yersinia pestis* that seemed comical at the time, but would have led to indictments and lawsuits today.[25] In 1940, Gross was a young researcher who worked briefly at the Pasteur Institute in Paris. There he immersed himself in the history of plague by reviewing the hand-written notes of Yersin and Simond, and meeting with scientists who were still actively engaged in plague research. Gross tells the story reported to him by a plague researcher named Dujardin-Beaumetz, who was studying the natural resistance to *Y. pestis* displayed by chickens, one of the few animals that cannot develop plague. One chicken he was studying was injected with enough plague bacillus to kill tens of thousands of people. It not only survived, but managed to escape through an open window. Dujardin-Beaumetz and his colleagues searched frantically for the bird, to no avail. A few days later, they learned that the chicken had been captured by a local building superintendent who took it home and had it roasted for a family dinner. Luckily, the microbe is heat-sensitive, and no one in that family got sick.

Like the great white shark, *Yersinia pestis* is a killing machine. It has one of the widest host ranges of any microbe, with the ability to kill more than 200 different animals, including the fleas that carry it. However, fleas are more resistant than mammals. They can survive for weeks harboring the bacterium, providing ample time to infect dozens of other animals and humans.

The plague bacillus's deadly program was developed through a series of genetic acquisitions it made sometime during the last 1,500 to 20,000 years. In 2003, *Yersinia pestis*'s entire genome was sequenced, more than four and a half million nucleotides in all, exposing the dark secrets of its success as a pathogen. The analysis revealed that it is derived from a cousin microbe, the

---

[25]Proceedings of the National Academy of Science, 92:7609–7611 (1995).

gastrointestinal pathogen *Yersinia pseudotuberculosis*, a fairly innocuous bacterium that causes a self-limiting attack of diarrhea. The *Yersinina pestis* genome like so many other bacterial species, is a mosaic derived from several different microbes, through an exchange of genes via plasmids, transposons, and bacteriophages (a process known as horizontal gene transfer-see chapter 1). The upshot of these genetic trades and acquisitions was the conversion of a fairly innocuous intestinal pathogen, spread by fecal-oral contamination, into a deadly bacterium transmitted by fleas into the blood and lymphatic systems of hundreds of mammals. Along the way, endogenous genes critical for gastrointestinal habitation that were present in the progenitor *Yersinia* species were lost or inactivated by mutations; these were unnecessary in the organism's new survival scheme and became a handicap, to be jettisoned or inactivated for the sake of genomic parsimony.

Critical virulence genes are found on three plasmids that make up about 3 to 4 percent of the bacterium's total DNA content. One of the plasmids is found in all three pathogenic strains of *Yersinia*, which include *Y. pestis* and *Y. pseudotuberculosis,* and another intestinal pathogen called *Y. enterocolitica*. But the other two plasmids–a small one, 9,600 nucleotides in length, and a giant one made up of more than 90,000 nucleotides–are unique to the plague bacillus. They contain the chemical codes for the pathogenicity genes that wiped out one-third of medieval Europe. Their analysis reveals the cleverness of *Y. pestis*'s two-pronged attack on the animal kingdom, via flea and mammal, with different sets of genes needed to provide the bacterium with the means to survive environments as diverse as the gastrointestinal tract of fleas and phagocytic cells of mammals.

One of the bacterial genes involved in both colonizing the flea and infecting mammals is *hms,* which utilizes a shrewd, if roundabout, method for transporting *Y. pestis* into mammalian blood streams. The plague bacillus forms a large colony in the flea foregut that plugs up the digestive tract. The bacterial plug prevents the fleas' food–mammalian blood–from being processed and digested. Without the means to properly digest a blood meal, infected fleas begin to starve. This makes them ravenously hungry, increasing their biting behavior. As fleas swallow blood from a healthy animal, the blood mixes with the bacteria in the obstructing plug. The flea then regurgitates some of the meal back into the host, along with a deadly load of bacteria. It is not exactly clear how *hms* promotes plug formation; scientists only know that *Y. pestis* bacilli that are engineered without it fail to obstruct the fleas' intestinal tract, thereby reducing their capacity to infect mammals.

Once inside a mammalian host, *Yersinia pestis* encounters its first obstacle, a microscopic blood clot formed at the site of the tiny flea bite, for which the gene *pla* (plasminogen activator) is used. Plasminogen is a blood protein that helps break down blood clots. A complicated but exquisite balance exists between two opposite yet intertwined biochemical processes–the timely formation and equally timely dissolution of blood clots. Without the proper balance, excessive bleeding would occur at one extreme, or dangerous clots would form inappropriately at the other. It all revolves around the synthesis and subsequent breakdown of the protein fibrin, nature's Band-Aid. Fibrin forms tight, interlocking bands that help seal holes in blood vessels caused by trauma. When an infected flea bites its mammalian host for a blood meal, a small blood clot forms at the puncture site which traps *Y. pestis.* Using *pla,* the bacterium dissolves the fibrin, allowing it to invade the circulatory and lymphatic systems.

The most elaborate genetic machinery involved in *Yersinia pestis* pathogenicity is reserved for evading its mammalian host's innate immune system. The plague bacillus is one of the fortunate microbes to have acquired the assemblage of genes needed to construct a type III secretion system, the protein syringe-and-needle apparatus used for transporting virulence factors into host cells (see chapter 1). The *Y. pestis* type III secretion system is used to disable host phagocytic cells, the first line of defense against an invading microbe. It does this by attaching to the cells and injecting about a dozen different proteins called Yops (*Yersinia* outer proteins). Two of these are YopE and YopT, which disrupt the cell cytoskeleton, preventing phagocytosis by crippling the arms (actually peudopods) of bacteria-eating white blood cells.

Without their supporting cytoskeleton, phagocytes are turned into useless blobs, like trying to fight in a boxing match with boneless limbs.

Another protein called YopH inhibits phagocytosis by blocking the signal generated when antibodies and complement bind to phagocytic cells, a signal that would ordinarily trigger bacterial engulfment. YopH actually orchestrates a two-pronged attack on the immune system. In addition to the action it takes against phagocytic white blood cells, YopH also inhibits lymphocytes by blocking their stimulation with cytokines and reducing the capacity for antigens to stimulate antibody production.

In addition, as if disabling phagocytes were not enough, another Yop protein, YopJ, stimulates premature apoptosis, cell suicide, which reduces the number of bacteria-eating cells available for defense.

In all, about 50 genes located on the large virulence plasmid orchestrate the construction of the type III secretion system and the delivery of a payload of Yops and supporting proteins. The resulting package is a molecular arsenal delivered into the very heart of the mammalian immune system, producing one of nature's most effective weapons of mass destruction.[26]

How did plague kill so many people so wantonly, yet leave others unharmed? Part of it, of course, is blind luck, depending on whether or not they were bitten by a flea or coughed on by someone with pneumonic plague. However, as described throughout this book genetic factors play an important role in surviving certain deadly infectious diseases–and the Black Death is probably no exception. The most intriguing survival theory is that the CCR5-Δ32 mutation protects against plague, an idea first proposed by Steven O'Brien and his colleagues at the National Institutes of Health (NIH). O'Brien was a member of one of the

---

[26]Detailed knowledge of the Yersinia pestis genome has provided scientists with the ability to distinguish different strains of the organism—a U.S. cultivated strain vs one from the former Soviet block, for example. In the event the plague bacillus should ever be used as a weapon of mass destruction by terrorists or other groups, some readers may find comfort in the fact that scientists will at least be able to identify the supplier.

research teams responsible for identifying the protective role that CCR5-Δ32 plays in slowing or stopping HIV progression. According to the NIH group's calculations, the CCR5-Δ32 mutation probably originated in a single individual about 700 years ago, and then expanded throughout Europe so effectively, because of disease resistance, that it is now found in tens of millions of people. O'Brien and his colleagues suggested that plague was a probable candidate for the expansion of this mutation, although other common infectious diseases including influenza, smallpox, typhoid fever, and tuberculosis were suggested as well. The fact that HIV and *Yersinia pestis* both attack the same target cells–macrophages–was used to support their hypothesis that plague was a likely possibility.

To make a case for an association between plague and CCR5-Δ32, O'Brien and his colleagues analyzed modern-day descendants of the survivors of the plague outbreak in Eyam, England, the self-sacrificing village (mentioned above) that was almost completely obliterated during the Great Plague. Many modern-day inhabitants of the village can trace their ancestry back to survivors of the plague outbreak. O'Brien reasoned that the mutation should be present in a sizable fraction of those residents. Indeed, CCR5-Δ32 was found in about 25 percent of them, which is a bit higher than its frequency throughout the rest of Britain, where about 20 percent of the population harbors one copy of the variant. The difference is modest, but it was enough to convince O'Brien and colleagues that they were on the right track. The hypothesis is intriguing: a mutation that enabled the population of its carriers to expand by Darwinian selection by providing them with a survival advantage against bubonic plague, a disease that generated a series of pandemics of historic proportions, that now functions in a similar way in carriers who are infected with the modern scourge HIV. Indeed, O'Brien's idea was so fascinating, it was chronicled in an excellent BBC documentary broadcast a few years ago. It's terrific drama. Unfortunately, the science is weak.

Recent investigations have punctured the hypothesis of a CCR5-Δ32/plague link full of holes. One of these is by a Danish group, who studied ancient bones that were at least four thousand years old; some samples dated back to the last Ice Age about

10,000 years ago. Genetic analysis of DNA extracted from the bones revealed that CCR5-Δ32 was highly prevalent in the sample, millennia prior to the Black Death and thousands of years before the predicted age of the mutation based on the NIH group's initial analysis. The conclusion: an epidemic disease may have selected for the variant in early inhabitants of Denmark, but it was probably not bubonic plague, which most likely entered Europe for the first time in the 14th century.

Another blow to the theory is a recent study by a group from Stanford using mice subjected to CCR5 "knockout" (that is, engineered to live their lives without any CCR5 genes), simulating homozygosity for CCR5-Δ32. To study the effects of *Yersinia pestis,* the Stanford researchers infected normal mice and the CCR5 "knockout" mice with equal loads of bacteria. It made no difference: both the engineered strain and the control mice were killed by the bacterium. There was no survival advantage provided by the loss of CCR5, at least in mice.

A third study, by Berkeley researchers Alison Galvani and Montgomery Slatkin, has also debunked the CCR5-Δ32/plague hypothesis. They used mathematical models to determine whether or not plague had the killing power necessary to cause rapid expansion of CCR5-Δ32 by Darwinian selection over a period of only four hundred years. Although there were two major plague outbreaks that had catastrophic effects on Europe, and many smaller epidemic eruptions before bubonic plague disappeared from Europe in 1750, Galvani and Slatkin's model suggested that the overall mortality rate was not sufficient to have raised the frequency of CCR5-Δ32 to more than a tenth of its current level. As far as being an engine for changing the human genome, *Yersinia pestis*'s stint as a pathogen was terrifying, but too brief, according to Galvani and Slatkin. In addition, plague started in Italy and spread north, whereas the frequency of CCR5-Δ32 is twice as high in Northern Europeans than in Mediterranean populations.

Galvani and Slatkin argue that the presence of this mutation increased in Europeans because it provided a Darwinian survival advantage against smallpox, not bubonic plague. Although plague killed a larger proportion of people than smallpox, the latter was a recurrent problem for a longer period of time. Its reign of terror lasted several thousand years before it was eliminated as a human pathogen only a quarter century ago.

Despite the flaws in the hypothesis that plague resistance is related to *CCR5-Δ32*, it does comply with the philosophy of J.B.S. Haldane (1892-1964), the British scientist, writer and Marxist philosopher, who believed that a scientific idea ought to be interesting even if it is not true.

CHAPTER 13

The King of Terrors

Smallpox is a uniquely human disease. There are no reservoirs of animals infected with the virus and no chronic, asymptomatic carriers. Infection with the smallpox virus leads to lifelong immunity. Thus, the only way to transmit the disease is from an infected individual to someone who lacks immunity. Since the smallpox virus can only strike once in any one person, it needs a large reservoir of people for its propagation and survival. Thus, it began to emerge as a major pathogen only after the great expansion in the human population that occurred after the development of agriculture. The first smallpox infection was probably caused by a virus that crossed the species barrier during an opportunistic encounter between an infected animal and a human being many centuries ago. The virus mutated in the progenitor human incubator to produce *Variola major,* the smallpox virus, which developed the capability of being transmitted from person to person. Monkeypox is a possible suspect for causing that initial animal-to-human infection.[27] Monkey-to-human transmission was also responsible for the development of the two major HIV strains that infect humans: HIV-1, found in the U.S. and Europe, which probably jumped from chimpanzees; and HIV-2, found in Africa, which appears to have

---

[27]In the spring of 2003, an outbreak of 71 monkeypox cases occurred in the Midwest, the first such infection in the United States. The monkeypox virus responsible for this outbreak was transmitted by exotic pets from Africa who were probably infected by monkeys before their capture.

originated in the sooty mangabey monkey. Both may have entered the human bloodstream during the butchering of infected monkeys. To put it bluntly, the modern world's worst infectious-disease pandemic–which has killed tens of millions of people, devastated several sub-Saharan nations, and changed human sexual mores around the world–may have been unleashed by some African tribes' penchant for eating monkey meat.[28]

The transformation of viruses that infect monkeys into ones that infect humans provided unique survival opportunities for both HIV and *Variola major*. Instead of propagating through simian species that number in the thousands or tens of thousands, the newly transformed viruses acquired the capacity to infect the mammal that has the largest population on the planet. In the course of smallpox's 10,000 year reign as a human pathogen, billions of human beings were infected. Its overall impact on humanity has been rivaled only by *Plasmodium falciparum*. During the past 20 years, HIV has infected hundreds of millions of people; and if an effective vaccine is not developed, AIDS may not disappear for centuries and billions will die.

Smallpox first appeared in Africa and south-central Asia and was probably brought to Europe during the Middle Ages by returning crusaders. The illness is characterized by very high fevers (up to 106 or 107 degrees Fahrenheit) and the eruption of ugly, marble-size, virus-filled pustules that can cover the entire body. The case-fatality rate–approximately 30 percent of adults and 80 percent of children–is one of the highest among all infectious diseases. The mortality rate was even higher in populations exposed to smallpox for the very first time. Survivors were often left with a lunar landscape of pockmarks on the face and body, resembling the pits formed from extreme cases of acne and chickenpox. The unsightly scars that marked smallpox survivors for life were a common sight in European cities and villages a few hundred years ago. In addition, as many as one-third of those who became infected were blinded, because the

---

[28]One of the arguments against using animal organs for human transplantation (the most common ones up to now usually being livers or kidneys from pigs) is that endogenous viruses may mutate into a human pathogen.

virus attacked the eyes. Smallpox is highly contagious, spreading from person to person by saliva from lesions in the mouth. The virus could be transmitted by coughing, kissing, sneezing, and even merely by talking face to face. The skin lesions are also highly contagious. The reward for surviving a smallpox infection was a lifetime of immunity from the disease.

The first recorded instance of an epidemic that is suspected of being smallpox occurred in 1350 BCE, during the Egyptian-Hittite war. It spread from Egyptian prisoners and quickly overran the Hittites. However, there is still some debate as to whether this epidemic was caused by bubonic plague, measles, or smallpox. It also appeared at about the same time in India and China. Evidence of smallpox infection has been found in several Egyptian mummies, including the remains of Ramses V, who died in 1157 BCE.

Smallpox changed world history by contributing to the downfall of several civilizations. The beginning of the downfall of the Roman empire coincided with a large-scale epidemic that began in about 180 CE, which killed between 3.5 million and 7 million citizens, including the Emperor Marcus Aurelius Antoninus.

Smallpox did not exist in the New World until it was brought to the Caribbean region by Spanish ships in 1507, only 15 years after Columbus's first voyage. It reached the American mainland in 1519, when Hernán Cortés led an expedition to the Aztec capital to conquer Mexico. The disease quickly overcame the Aztecs, who were highly vulnerable to the new infection; the Mayan and Inca empires also succumbed. Nearly half the indigenous population of Mexico and Central America died of the disease within a few years. With the aid of the smallpox virus, contingents of Spanish conquistadors numbering only a few hundred men were able to overcome three of the world's great empires, with a combined population in the tens of millions. It was perhaps the greatest military upset in human history, equivalent to a small militia successfully taking over the state of California.

The terrible disease carried by the Spanish contributed to the radical change that occurred in New World demographics. In 1518, the population of Mexico was approximately 25 million; by 1620, in the aftermath of smallpox, warfare, genocide, measles, and other diseases, only 1.6 million remained. On the island of Hispaniola (now occupied by the nations of Haiti and the Dominican Republic), the entire indigenous population of about 2.5 million was wiped out within a decade of Spanish occupation, largely from smallpox. The devastation caused a severe labor shortage on plantations growing sugar cane and prompted the mass importation of African slaves.

The flow of infectious disease was not just from Europeans to indigenous people. It is believed that the crew of Christopher Columbus's ships picked up syphilis (the "Great Pox") from the New World and brought it back to Europe. It was first described in Spain soon after Columbus returned from his last voyage, by Spanish physician Ruy Diaz de Isla, who asserted that he treated some of Columbus's crew members for the newly discovered disease. Syphilis would eventually infect between 8 and 14 percent of Europeans by the beginning of the 20th century and become the leading cause of debilitating heart and neurological disease.

Smallpox was the most dreaded disease in colonial America, repeatedly striking every major city. The first known case there appeared in the Massachusetts Bay Colony in 1616, where it had an effect on native populations similar to what happened in Mexico and Central America. About 90 percent of the Algonquin tribe living in Massachusetts at the time perished, clearing the way for the Plymouth settlement a few years later. The 1616 outbreak was the first of dozens that would affect Massachusetts in the next two hundred years as new immigrants brought fresh virus from Europe. The epidemic in Boston in 1752 was particularly severe, infecting more than one-third of the population.

In the 1700s, nearly 10 percent of the population of England died of smallpox; approximately one-third were pockmarked. It remained endemic in Europe until the 19th century and was responsible for one out of every 10 deaths. Nearly half a million Europeans died every year, affecting rich and poor alike, including many members of Europe's royal families. In England, Queen Elizabeth I had a nearly fatal case in 1562, and Queen Mary died of it in 1694, as did her son the Duke of Gloucestershire, the only direct heir to the throne, in 1700. His death left a void in the monarchy that was eventually filled by a distant German relative from the House of Hanover who became King George I, and the Hanovers have maintained their hold on the British monarchy ever since.[29] In all, four reigning European monarchs died of smallpox in the 18th century.

The armies of Napoleon and of George Washington were nearly wiped out by the disease. During America's Revolutionary War, more than half of Washington's 8,000 troops contracted smallpox, preventing a planned invasion of British troops stationed in Quebec. The British soldiers were more resistant to the disease because many of them had been vaccinated by a method known as variolation (see below).

Like plague, smallpox was also used as an early bio-weapon. During the 14th century, invading Tartars catapulted smallpox-laden corpses over the walls of besieged towns. Sir Jeffrey Amherst, commander-in-chief of British forces in North America during the French and Indian Wars (1754-67) suggested that blankets and handkerchiefs contaminated with the scabs of smallpox victims be sent to the Indians–a desperate attempt to cut off their successful raids on forts and settlers. Although there is no direct evidence that this plan was ever carried out, a smallpox epidemic nearly wiped out Native American tribes in Ohio soon thereafter and helped put an end to the bloody Indian uprising there. Smallpox also had a hand in decimating the Huron and Iroquois nations, and later, several tribes in the Southwest. The disease carried by Europeans proved to be a more effective weapon than firearms in helping to overcome Indian resistance to the continuous spread of settlers throughout the American West.

Smallpox was the first disease to be controlled by immunization. Astute observers in ancient China, Africa and India noticed that survivors of smallpox were resistant to later outbreaks. Similar observations were made by Thucydides, an Athenian aristocrat who wrote the history of the Peloponnesian War in 430 BCE. The Chinese, in about 1700 BCE, developed an oral vaccination using pills made from fleas removed from cows infected with cowpox. Dhanwantari, a Hindu physician who is

---

[29]Members of the House of Hanover changed their name to House of Windsor during World War I to hide their German heritage.

thought to have lived about 1500 BCE, also described the protective value of cowpox. Centuries later, these lessons were learned all over again, leading to the direct inoculation of cowpox material by Jenner beginning in 1796, a procedure that revolutionized the prevention of smallpox in the 19th century (see below). In about 600 BCE, the Chinese developed a method of inhaling powder from dried smallpox scabs through the nose. These observations ultimately led to the practice of inoculating small quantities of infectious material, obtained from the pus and scabs of people with smallpox, into an open incision of the arm of a recipient; a technique known as variolation (after *variola*, the Latin word for smallpox). The goal was to induce a controlled mild infection that would provide lifelong protection against natural outbreaks of the deadly disease.

Variolation originated in India and China but it was perfected by physicians in the Ottoman Empire. The practice spread throughout Europe after it was introduced by Lady Mary Wortley, an English aristocrat who learned about the technique in Turkey, where her husband served as the British ambassador. Her face had been disfigured with pockmarks caused by a smallpox attack when she was a young woman, and the same outbreak killed her 20 year old brother. She became a passionate advocate of variolation after her return to England; in fact, her daughter was the first person in England to be variolated.

During the early 1700s, human trials conducted in Britain on condemned prisoners and on orphans demonstrated the effectiveness of variolation. However, the practice was associated with a high complication rate. Variolation was intended to cause a mild case of smallpox confined to the site of inoculation, but full-blown, fatal infections sometimes developed. It could also cause a severe local reaction, initiated by a powerful immune/inflammatory response, capable of destroying the inoculated limb (pox viruses are some of the most potent inducers of the immune system). The lesions themselves could transmit smallpox to others, and other infections were sometimes transmitted from the donor of the smallpox material to the recipient, the most common being bacterial pathogens of the skin and syphilis. When performed by some practitioners, the practice had a rate of severe complications and mortality as high as 2 to 3 percent, equivalent to the mortality rates associated today with coronary artery bypass surgery. Yet, despite the hazards, variolation was far safer than a naturally acquired smallpox infection. Consequently, the technique spread throughout Europe.

Variolation was introduced to the American colonies by Dr. Zabdiel Boylston (John Adams's great-uncle) and by Cotton Mather (who, as a 12 year old, studied at Harvard College, and later became a famous Puritan minister, author, and amateur scientist). Mather learned about variolation from an African slave named Onesimus, who informed the minister that the procedure was common in Africa and that he had been inoculated by his own mother: "My mother scratched my skin and I got sick, but lived to come here, free of smallpox, as your slave," he told Mather.[30]

Because of the risk, it took years before variolation was accepted in the American colonies. During the smallpox epidemic in Boston in 1721, Mather inoculated his young son, who nearly died from the treatment. Many Bostonians believed that variolation was a plot to spread smallpox. Mather's advocacy for the controversial treatment resulted in his house being bombed. Eventually, though, as smallpox outbreaks erupted throughout the colonies, the procedure caught on. In many studies, which, at the time, were the earliest large-scale clinical trials of a medical treatment ever conducted, variolation proved to be very effective. During the epidemic that hit Boston in 1753-1754, 6 of 244 people who were variolated by Boylston contracted smallpox, compared with 844 of 5,980 who had not been treated. People who could afford the treatment eventually flocked to doctors, willing to accept the risk and weeks of illness, confinement, and recuperation. Adams and his entire family were inoculated just before the onset of the Revolutionary War. Adams was no stranger to the dangers of infectious disease: his father died of epidemic influenza and his brother Elihu died in 1775 of the "bloody flux," epidemic dysentery, while serving as a captain in the militia. During his variolation confinement, Adams sent numerous letters home to his wife Abigail. Fearing that he might contaminate other members of the

---

[30]Quoted from Susan Donnelly in "Inoculation," *The New Yorker*, Aug. 4, 2003, p. 38.

household, Adams also sent along instructions to have the letters burned soon after they were read. In 1736, Benjamin Franklin's four-year-old son Francis died during an epidemic, not having been inoculated–a decision that the great scientist and statesman regretted for the rest of his own life.

Prevention of smallpox was transformed in the late eighteenth century by Jenner, a physician and variolation practitioner who heard local tales about milkmaids bragging that they were resistant to smallpox. The milkmaids claimed that they became resistant due to their contact with the udders of cows infected with cowpox, which gave them a mild pox infection on their hands. One of them told Jenner during a consultation, "I shall never take the smallpox for I have had the cow-pox." Because of their apparent immunity to smallpox, milkmaids were often asked to help care for children afflicted with the disease. Indeed, Jenner noticed that milkmaids rarely had pockmarks on their faces, the all-too-common visible reminder of a past infection. He also observed that people who had a history of cowpox did not form lesions when he variolated them. By 1780, Jenner was already telling colleagues about the possibilities of preventing smallpox by injecting people with cowpox. He presented the idea at medical societies, where he was viewed as a nuisance and threatened with expulsion. However, he had Devine faith in the idea of finding a benign prophylaxis for smallpox: "I have sometimes found myself in a kind of reverie...; it is pleasant to recollect that those reflections always ended in devout acknowledgment to that Being from whom this and all other blessings flow."

To prove that cowpox prevented smallpox, Jenner conceived a clever but unethical experiment (by today's standards), using an eight-year-old boy named James Phipps as a test subject. On May 14, 1796, he prepared a cowpox extract from the hands of a milkmaid named Sarah Holmes, inoculated the boy, and then attempted to variolate him with the standard method. When non-immune individuals are variolated, a pox lesion develops at the inoculation site. People who survive a smallpox attack do not respond to variolation, because their immunity prevents the local infection from taking hold. Similarly, Jenner's test child was resistant to variolation; the cowpox had induced smallpox resistance, Jenner reasoned. Jenner named his technique *variolae vaccinae* ("smallpox of the cow"; vacca is latin for cow), which was later Anglicized as "vaccination." At the end of 1796, Jenner sent an article to the Royal Society describing thirteen people who had previously had cowpox and in whom variolation had induced no reaction. It also described the experiment with James Phipps. However, Sir Joseph Banks, the President of the Royal Society, and Sir Everard Home rejected the manuscript for publication in Philosophical Transactions of the Royal Society. The Council of the Royal Society repulsed Jenner because he was "in variance with established knowledge" and "incredible." Jenner was further warned: "He had better not promulgate such a wild idea if he valued his reputation."

Jenner expanded his tests to include a larger number of subjects, and similar results were obtained. His results ultimately proved to be so convincing, and his method of prophylaxis was so much safer than variolation, that cowpox vaccination spread throughout Europe and the United States within a few years. Thomas Jefferson had himself and his entire household vaccinated, and learned how to administer the treatment himself. Jefferson praised Jenner in a letter, stating that "future generations will know by history only that the loathsome smallpox has existed, and by you had been extirpated."[31] Jefferson had some federal funds set aside to create the first government-sponsored vaccination program. In France, Napoleon had his troops vaccinated and was so pleased with the results that he pardoned British prisoners when Jenner intervened on their behalf.

Because natural cowpox infection in cows was so sporadic, Jenner eventually developed the "arm-to-arm" procedure, in which fresh cowpox virus was passed directly from person to person, circumventing the need for a bovine intermediary.

The effectiveness of Jenner's treatment was further established in studies conducted elsewhere in Europe and the United States. In one study, carried out in 1835 during an outbreak in Marseille, the death rate was 12.4 percent in the unvaccinated group, but less than 1 percent in those treated with the Jenner

---

[31]Rene J. Dubois, *Louis Pasteur: Free Lance of Science* (Boston: Little, Brown, 1950).

method. Vaccination studies in the 19th century were the first to use the new science of statistical analysis to inform physicians and the public about a health issue. As a result of Jenner's treatment, variolation became obsolete, and laws were eventually passed in European parliaments banning its use. But even cowpox vaccination was not without some risks, occasionally producing serious, life-threatening infections. It was, after all, a live virus vaccine. Very young children, and people in poor health or with defective immune systems would sometimes develop a severe, potentially fatal systemic infection called disseminated vaccinia or the cowpox wound would become infected with skin bacteria, or blood-borne pathogens from the donor, as mentioned earlier in this chapter. However, the overall risk was minimal compared to the chance of contracting smallpox. Even so, despite the very favorable risk/benefit ratio, many people still objected to compulsory vaccination.

Opposition to vaccines accelerated after legislation was passed by the British Parliament in 1853 and 1861 that made smallpox vaccination compulsory. Violators of these laws were punished with a fine or brief imprisonment. Those who opposed the laws were outraged that a person could go to prison for essentially doing nothing, and they objected to vaccinating children against the will of parents who believed that they were protecting them from a dangerous infection that was sure to result from the vaccine's side effects. To opponents of the vaccine, the British government was infringing on the rights of individuals to make personal decisions about themselves and their children. Those imprisoned for refusing smallpox vaccination were viewed as martyrs.

One of the most vocal critics of vaccination was Alfred Russel Wallace (1823-1913), a naturalist, writer, and social commentator who, along with Darwin, developed the principle that new species evolved through survival of the fittest. Wallace formulated his ideas during the decade he spent in the Malay Archipelago collecting thousands of plant and animal samples. He was an excellent observer of nature and was the first to recognize the distinct differences in the flora and fauna of Australia compared with Asia, suggesting separate evolutionary pathways caused by the island-continent's geographic isolation. In 1858, Wallace was stricken with malaria. While suffering from the intermittent fevers characteristic of that disease, he was unable to work and was forced to lie in bed and ruminate. As Wallace later recalled, "every day during the cold and succeeding hot fits I had to lie down for several hours, during which time I had nothing to do but think over subjects that particularly interested me." During one fever attack, he had an epiphany. Thinking about the work of Thomas Robert Malthus (1766-1834)–who had written in *An Essay on the Principle of Population, as It Affects the Future Improvement of Society* that disease, accidents, war, and famine placed a check on population expansion–Wallace realized that the strongest, fastest, and most cunning survived to reproduce, improving the fitness of each generation: "Why do some live and some die? On the whole, the best fitted live." Wallace waited impatiently for the fever attacks to subside so that he could put his ideas into words, and then spent the next two days writing a report, which he quickly sent off to Darwin. Darwin was quite distressed when he received the paper, as he had been formulating precisely the same ideas over the past two decades and was in the process of writing "On the Origin of Species," his great tome on the subject of evolution, a project that only a few close friends in the sciences were aware of. After spending nearly 20 years deliberating over the concepts of evolution and survival of the fittest, he was about to be scooped. Taking the advice of his close associates, Darwin wrote an abstract of his book, which would be published together with Wallace's theory. The experience forced the ever-procrastinating Darwin to finally complete "On the Origin of Species." This monumental book, which forever changed our fundamental ideas about life on Earth, was published about a year and a half later. In the introduction to his own book, Darwin generously acknowledged Wallace's contribution.

Wallace staunchly opposed compulsory vaccination. He accused the medical establishment of misrepresenting the facts regarding the Jenner vaccine by overstating its prophylactic capacity and understating its side effects. He argued that the original experiments performed by Jenner were flawed because of the absence of a control group. A modern version of the cowpox experiment would have included a control group treated with a sham vaccine to determine the baseline percentage of people in the study population who do not form a variolation lesion (previous smallpox infections, poor variolation technique, or weakened inoculation material could result in the absence of a lesion). Wallace also used epidemiological data, which included comparisons of large groups of vaccinated and unvaccinated people in towns and villages in the United Kingdom and members of the British Army and Navy, to argue that the Jenner vaccine was not only useless, it was downright dangerous, especially in children. He accused the medical establishment of promoting a dangerous procedure for their own financial gain, and he condemned government officials and the British Parliament responsible for enacting compulsory vaccination legislation, declaring them gullible and ignorant of what Wallace regarded as the facts in the matter. In an essay articulating his anti-vaccination beliefs, published in 1898, Wallace wrote,

"And when we consider that these misstatements, and concealments, and denials of injury, have been going on throughout the whole of the century; that penal legislation has been founded on them; that homes of the poor have been broken up; that thousands have been harried by police and magistrates, have been imprisoned and treated in every way as felons (for failing to vaccinate their children thereby breaking the law); and that, at the rate now officially admitted, a thousand children have been certainly killed by vaccination during the last twenty years, and an unknown but probably much larger injured for life, we are driven to the conclusion those responsible for these reckless misstatements and their terrible results have, thoughtlessly and ignorantly but none the less certainly, been guilty of a crime against liberty, against health, and against humanity, which will, before many years have passed, be universally held to be one of the foulest blots on the civilization of the nineteenth century."

Two things—the historic eradication of smallpox using a modern version of the Jenner vaccine, and an unbiased analysis of the epidemiological data—ultimately proved that Wallace's point of view was wrong. His rabid anti-vaccination sentiment along with his belief in spirituality and phrenology (the notion that personality traits and intelligence can be discerned from variations in the shape of the head) damaged Wallace's scientific reputation, unfairly diminishing his role as a co-discoverer of the survival of the fittest principle.

The anti-vaccination movement, which had supporters in the medical establishment and among members of Parliament, was eventually successful in getting Parliament to modify the compulsory vaccination laws. In 1907, it passed a law that gave parents the right to refuse vaccinating their children if they had strong religious or philosophical objections. It was the first of many skirmishes in the battle between public-health policy and personal medical decision-making that continues to this day.

In 1939, a British scientist named Allan Downie showed that the cowpox virus that had been used to vaccinate hundreds of millions of people all over the world had spontaneously changed to a less infectious virus at some time over its continuous century and a half of use. He named the variant viral strain *Vaccinia*. No one knows when the original cowpox virus changed to *Vaccinia,* but it became the strain most people now over the age of forty received as children, when routine smallpox vaccination was still in practice. Individuals who were successfully vaccinated as children have a small, dime-shaped scar on their upper arm as an aftermath.[32] Safe transit in humans was the Darwinian strategy used by cowpox virus in its conversion to *Vaccinia*. Vaccinators apparently had inadvertently selected a milder strain of virus by continuous person-to-person transfer, using viral-laden material from donors that caused fewer side effects and smaller pox lesions. *Vaccinia*'s survival strategy was to make itself so innocuous, it could be given to a child.

As a result of vaccination and strict isolation of infected cases, smallpox declined dramatically in the United States between 1900 and 1949. During this period, there were "only" 16,860 casualties. The last case of smallpox reported in United States was in 1949. Mandatory vaccination was eventually terminated in 1971; the vaccine itself has a mortality of one per million, so its usefulness was far outweighed by the risk.

---

[32] I don't have a scar because my pediatrician (whom I still recall with fondness and awe), seems to have botched my smallpox inoculation.

However, smallpox maintained its position as one of the top two or three infectious disease killers in poor countries where vaccination programs were non-existent. Even after the disease had been wiped out in the United States and Europe, it continued to take several million lives a year throughout the rest of the world. Yet, the successful eradication of smallpox in the developed nations gave public-health officials hope that stepped-up vaccination programs–and political will–would do the same in those other countries.

A widespread campaign to eliminate smallpox was first conceived of by a 19th century monarch, King Charles IV of Spain, an early advocate of the Jenner treatment. The king had his entire household vaccinated after his daughter, Princess Maria Luisa, suffered a near-fatal smallpox attack. His interest in vaccination grew with his concern over the devastating effect that smallpox was having on the Spanish colonies. He decided to send a contingent of physicians and nurses to vaccinate people in the hardest-hit areas, and chose Francisco Xavier de Balmis to lead the expedition. Balmis came up with a plan to maintain a fresh supply of cowpox virus on his trans-Atlantic journey, by bringing along two dozen orphans to serve as incubators to maintain a continuous line of fresh cowpox virus using person-to-person transfer. He vaccinated a pair of the orphans just before the voyage set out and used them to transfer the virus to a few of the others during the voyage to the Spanish colonies, who then brewed the virus for the next pair, and so on. Stocks of virus were also maintained by storing cowpox secretions between two glass slides sealed with paraffin; but this was not as effective as using live cowpox virus scraped from lesions of the recently inoculated. Since the discovery of viruses was decades in the future, Balmis's use of living cultures and stored infectious material was based entirely on empirical observations and trial and error. For three years, Balmis traveled throughout the Americas, including Texas and California, and across the Pacific to Canton and Manila, teaching his technique to local doctors and vaccinating volunteers. Starting with his initial pair of inoculated orphans, nearly one hundred thousand people in the Spanish colonies were ultimately vaccinated.

The modern worldwide smallpox eradication program was initiated in 1967 by the World Health Organization. We do not know who the first human was to contract smallpox, or the names of billions more who have been stricken over the millennia (with some notable exceptions). However, we do know the last confirmed case in the world: Ali Maow Maalin, a Somali cook, who contracted the disease on October 26, 1977. For the first time in history, an infectious pathogen was deliberately and successfully forced to extinction—the one and only species loss that humanity can view with pride.

Smallpox virus now exists as a vestige of Cold War paranoia, in vials of frozen stock cultures hibernating in the freezers of government-regulated labs in the United States and in the former Soviet Union. More than a hundred tons of virus-enriched solutions were generated as part of these two superpowers' germ-warfare programs. In addition, Iraq, Iran, India, Pakistan, Israel, China, and Korea might also have stockpiles of *Variola major* (although it should be noted that no hidden stocks of smallpox virus were found in Iraq after the U.S. invasion in 2003). During the late 1980s, Soviet and American scientists debated whether or not existing *Variola* stocks should be destroyed. The pro-destruction argument was based on a fear that theft by terrorists or a lab accident might result in release of the virus. The anti-destruction side felt that there were enough safeguards in place to prevent a catastrophic release, and that, in the event that there were hidden stocks elsewhere in the world, the scientific world had to be prepared to study the virus in the laboratory. The anti-destruction argument won out in the end. In 1995, the World Health Organization brought the issue back for discussion and proposed a four-year timetable during which all countries would destroy their hidden supplies of *Variola* virus, a deadline that was extended several times. But after the terrorist attacks on September 11, 2001 in New York City and Washington, D.C., and, a month later, the mailing of anthrax spores to several government offices and news agencies, the notion that a terrorist group might one day launch a smallpox attack was suddenly regarded as a real possibility. The frozen stocks of smallpox will have to stay around a little longer in the event that researchers need to pursue further laboratory investigation of the virus.

With the eradication of smallpox, widespread vaccination against the disease ended in the late 1970s. However, interest in the *Vaccinia* vaccine has reemerged because of the bioterrorism threat, and it is being offered to health-care professionals and armed-forces personnel who would be among the first to become exposed to smallpox virus in such an attack. More than 40 percent of the people in the United States were born after mandatory smallpox vaccination ceased, and the immune status of those who were vaccinated four or five decades ago is uneven. A new smallpox outbreak launched by bioterrorists on a non-immune public could be devastating. However, the vaccine's one-in-a-million mortality rate and the serious complications it causes in a substantial minority of recipients (including myocarditis, an inflammation of heart muscle) have tempered enthusiasm for its widespread use at this time.

No other research endeavor (aside from astronomical observations) connects the ancient and modern scientific worlds more than the search for effective and safe ways to curtail the disease that was dubbed "the king of terrors" by Dr. Benjamin Rush, a noted Philadelphia physician and Revolutionary War patriot. Although there have been no new cases during the past quarter century, smallpox vaccination research is still quite vibrant, focusing on the search for a safer alternative to *Vaccinia* for widespread use in the event of a smallpox resurgence. Also, the powerful effect that pox viruses have on the immune system is being exploited to develop vaccines against microbes that do not generate an adequate immune response on their own. Recombinant DNA is being used to team up *Vaccinia* with weaker immune stimulators to create more potent vaccination recipes (see chapter 14). The first "recombinant vaccines" containing *Vaccinia* were developed in the 1980s as potential measures to fight influenza. Although they were not used in clinical practice, they demonstrated what scientists call "proof of principle," a concept that turns out to work experimentally–in the laboratory, if not in clinical practice. In 2003, several groups developed recombinant vaccines containing various HIV genes inserted as passengers in the *Vaccinia* genome. Thanks to the stimulating power of pox viruses, many of the recombinant vaccines evoked a strong immune response against HIV under laboratory conditions. Clinical trials will likely follow for at least a few of these creative vaccines—marriages of two viruses that have played such a significant role in human history, made possible by modern molecular genetics. The persistent scientific interest in what should by now be a dead disease is testimony to the power of pox viruses to stimulate our immune systems, and our anxiety levels.

Galvani and Slatkin have argued that the CCR5-Δ32 mutation expanded in northern Europeans by protecting against smallpox rather than bubonic plague (see chapter 12). Smallpox has approximately half the case fatality rate that plague has, but it infected humans for a much greater period of time. Cases began to appear during antiquity, and continuous epidemic waves pummeled European and other populations until the middle of the 20th century. Taking into account smallpox's mortality rate and its more enduring effect on the population, it would have increased the frequency of CCR5-Δ32 to current levels, assuming a 50 percent protective rate, according to calculations made by Galvani and Slatkin.

In addition to their mathematical models, there are biochemical and molecular arguments favoring a link between CCR5-Δ32 and smallpox. Like HIV, pox viruses use CCR5 receptors as gateways to host cells. One is the myxoma virus, which causes a lethal immune disorder in rabbits. It can be made to infect rodent cells if CCR5 receptors are expressed on the surface. Conversely, if access to CCR5 receptors is blocked with antibodies, myxoma virus fails to infect the cells. Whether this is also true for *Variola major* is not known at this time. Indeed, it is a question that may remain unanswered. Although vaccine research is very active, there is not much incentive to explore the basic molecular biology of a disease that no longer exists. Such experiments are too dangerous to perform using live viruses. Even the most stringent isolation procedures cannot completely safeguard research on deadly organisms, as evidenced by the number of laboratory workers who have died while working on Ebola virus and other emerging pathogens. Restoring frozen stocks of smallpox virus back to life would also create a risk to the public. An alternative method for studying interactions between *Variola major* and CCR5 receptors would be to use non-infectious proteins from the virus generated by recombinant DNA. However, a negative result might be difficult to interpret.

So, determining whether or not the CCR5-Δ32 gene variant influenced the course of smallpox may remain an enigma for some time. It may be unscientific to admit, but, considering the impact smallpox has had on human history, it might not be too terrible if the "king of terrors" remained buried with a few secrets.

CHAPTER 14

Gene Splicers

Humans first learned how to exploit the microbial world thousands of years ago through trial and error, by discovering fermentation. Because of the ubiquity of microbes capable of fermenting sugars, nearly every culture throughout history inadvertently stumbled upon recipes for converting sugar from local fruits and vegetables into alcohol. Science only began to play a role in this ancient practice after Pasteur's discoveries; the art of alcohol production improved after he found that fermentation was caused by living microbes. With inoculation against smallpox–first using pus from pox lesions, and then Jenner's cowpox–humans began to harness microbes as weapons in the battle against infectious disease. Again, it was Pasteur who transformed what was then a rudimentary practice by bringing it into the modern world, turning vaccine development into a scientific enterprise. With his deliberate manipulation of bacteria and viruses, Pasteur was able to take infectious organisms and strip them of their virulence without disturbing their capacity to stimulate an immune response. This basic strategy for developing vaccines continues to this day.

The ability to harness microbes' power relied on choosing microorganisms that had altered their genome in some manner to provide traits that could be exploited for humankind's benefit. Before the modern era, such alterations were all accidents of nature that crossed paths with observant eyes. But with modern science, a new phase evolved in our ability to exploit the microbial world, based on the discovery–actually, the invention–of recombinant DNA. This phase began more than 30 years ago when scientists developed a method for mobilizing plasmids and bacteriophages, the genetic vehicles that infectious microbes use to transfer antibiotic resistance and virulence factors, in order to create novel gene combinations that would ordinarily not occur in nature. The method involves cutting and pasting together pieces of DNA to generate recombined (usually called recombinant) genes, analogous to manipulating images and text from one computer program to another. Genes pasted onto plasmids and bacteriophage can be transferred into bacteria, where they propagate, turning the host microbes into gene factories. By growing recombinant genes–the process is called molecular cloning–scientists are able to harvest sufficiently large quantities of genetic material to determine a gene's DNA sequence; and, with simple modifications, recombinant genes can also be activated in microbes, forcing them to produce large quantities of proteins encoded by recombinant genes, essentially providing an unlimited source of protein in amounts not possible, or economically feasible, with any other manufacturing process. The proteins can be easily purified for scientific purposes and for treating disease, such as making human insulin for diabetes.

Through recombinant DNA, tens of thousands of individual microbial genes have been isolated and analyzed. The entire genome of scores of infectious agents (as well as those of their insect vectors) has been sequenced, providing a catalogue of genes that scientists can exploit to create new therapies and diagnostics. We can now diagnose many infectious diseases by examining their DNA using PCR-amplification techniques, which takes only a few hours, instead of culturing them, which takes days and weeks. This is especially important in dealing with slow-growing organisms such as *Mycobacterium tuberculosis*, and for diagnosing certain viral infections. Humans are battling back with the very weapons that microbes have used against us. On a grander scale, recombinant DNA has provided researchers with the tools enabling us to determine the entire sequence of the human genome and to understand the causes of cancer, genetic disorders, and other conditions.

The recipe for growing recombinant genes was derived entirely from components commandeered from the microbial world: enzymes to cleave or cut DNA; enzymes to paste different combinations back together again, to each other, and to plasmids and bacteriophages; microbes to receive and replicate DNA; and a Darwinian selection strategy to identify bacterial clones containing the recombinant genes.

The primary DNA cleavage enzymes used in recombinant DNA technology are known as restriction endonucleases or restriction enzymes, which cut double-stranded DNA at very specific sites, providing researchers with small, workable fragments of DNA that are uniform and predictable in size. Restriction enzymes bind to short stretches of DNA and cut across the DNA molecule's double-stranded backbone. Each restriction enzyme has its own recognition pattern. For example, the restriction enzyme called *EcoR1* binds to the double-stranded DNA sequence:

GAATTC

CTTAAG

Wherever this signature sequence of consecutive nucleotides occurs in the genome, *EcoR1* will attach and cut DNA at that site. Imagine two long interlocking pieces of white string (representing a segment of double-stranded DNA) with random dots of black (*EcoR1* recognition sites) and a pair of precision scissors (*EcoR1*) that can only cut the string where the black dots occur. There are approximately one million *EcoR1* binding sites in the human genome, but only a few such sites, and sometimes only a single site, in simple structures such as bacteriophages and plasmids, which have relatively small genomes. The marvelous property of restriction enzymes as a research tool is their ability to cut DNA from any source, whether it's extracted from human cells or from the most primitive viruses.

Restriction enzymes are unique to bacteria; they do not exist in higher life forms. The first such enzyme was isolated in the late 1960s by Hamilton O. Smith, a researcher from Johns Hopkins, who was investigating the host-range specificity of bacteriophage– that is, he was trying to discover why bacteriophages active in one bacterial species are inactive in others. He and his colleagues were studying the bacteriophage P22, which easily infects and destroys *Salmonella* but fails to subdue *Hemophilus influenzae*, a bacterium that is a common cause of ear and sinus infections, and meningitis. Smith found that the bacterium was able to avoid infection by disrupting P22's genes using a specific enzyme, which he later isolated and named *Hind II*–it was the first restriction endonuclease ever found. Others were soon identified in a host of bacteriophage-resistant bacteria. Smith had uncovered the bacterial world's version of an immune system, a molecular scheme for disabling invading viruses by applying a karate chop to the intruder's genome.[33]

Bacteria that produce restriction enzymes put themselves in a bit of a chemical dilemma. Since restriction endonucleases cleave DNA regardless of origin, without discrimination, how do they avoid turning against themselves by cutting up their own genome? It turns out that bacteria capable of producing restriction enzymes have developed a clever strategy of deception to avoid genetic self-destruction: they protect their genome by producing other enzymes that chemically modify restriction-enzyme recognition sites, preventing DNA cleavage. The modification is the addition of an organic molecule called methyl, which is simply a carbon atom attached to three hydrogen atoms. Enzymes called DNA methylases add methyl groups to adenine and cytosine nucleotide bases located on restriction-enzyme recognition sites, rendering them impervious to endonuclease digestion. In higher animals, methyl groups that are used for a variety of biochemical purposes are generated via a metabolic pathway involving vitamin B12 and folic acid. Humans and other eukaryotes also methylate DNA, at certain cytosine bases, but not to protect against the digestion of restriction endonuclease, since we do not synthesize these enzymes. Instead, it is used as a tool for regulating gene

---

[33]Smith shared a Nobel Prize in Physiology or Medicine in 1978 with Daniel Nathans and Werner Arber—Arber for his work on host-range specificity in *E. coli* and discovering a DNA cleavage enzyme in that bacterium that disrupted the genome of an invading bacteriophage, and Nathans for pioneering the application of restriction enzymes to problems of genetics during the 1970s, using the enzymes to dissect viral genomes as a way to help understand the function of individual viral genes.

expression by preventing transcription factors from binding to gene promoters. In general, the more a gene is methylated, the less it is expressed. DNA methylation is an important factor in regulating gene expression in a tissue specific manner-genes that code for proteins involved in transmitting signals through the brain are less methylated in the neurons than they are in skin cells, for example. Without DNA methylation, gene expression is chaotic and mindless. Loss of controlled DNA methylation is one of the many reasons that gene expression in cancer cells can be so disordered that hair and teeth can sometimes grow out of certain ovarian tumors, and lung cancers sometimes produce toxic amounts of a hormone normally secreted by the brain's pituitary gland.

Modifications of DNA that alter its regulation without affecting its primary nucleotide sequence is referred to as epigenetics. The control of gene expression through epigenetics is now recognized as a key element in deciphering the genetic code; almost as important as the information contained within the primary sequence of an organism's genome.

DNA methylation plays an important role in embryonic development. Some genes are methylated in one germ cell and remain unexpressed even after fertilization; the only functioning copy is the one contributed by its unmethylated counterpart in the other germ cell. This pattern of selective gene expression from a single parental source is called imprinting. Imprinting makes a difference for some genetic disorders. For example, women with Turner's syndrome, a chromosomal disorder caused by inheriting only a single X-chromosome instead of two, will have slightly different manifestations depending on whether they have retained the maternal or paternal X-chromosome, because a few X-linked genes are imprinted. Women with Turner's syndrome who retain only the paternal X-chromosome tend to demonstrate better verbal skills and social functioning compared with their genetic compatriots who retain the maternal X-chromosome. One of the more dramatic examples of the effects of imprinting in genetic disease involves a pair of rare disorders called Angelman syndrome and Prader-Willi syndrome, both caused by abnormalities in an imprinted locus on chromosome 15. If an individual's maternal transmitted imprinted locus is damaged, he or she will develop Angelman syndrome, which is characterized by an odd constellation of symptoms that include a persistently happy disposition (with frequent laughter), hyperactivity, learning problems, and unusual physical movements and seizures. Damage to the paternal transmitted region leads to Prader-Willi syndrome and its equally odd symptom profile of insatiable appetite, obsessive-compulsive disorder, temper tantrums, morbid obesity, and short stature.

Although there are differences in how bacteria and humans use DNA methylation to control genes, one function is shared by both. Humans also use it as a strategy to disable invading viruses. The retroviruses and transposons that have invaded the human genome in the course of evolution have been suppressed for eons by DNA methylation, which squelches their expression. The powerful stranglehold that DNA methylation has on gene expression is not only passed down from parent to child, in the example of imprinted genes, but, under some circumstances, is sustained for the duration of a species' existence.

Most restriction enzymes cleave double-stranded DNA asymmetrically, leaving several single-stranded nucleotides free on the ends. These ends hook up easily with other DNA molecules that have been digested by the same enzyme, like adjacent pieces in a jigsaw puzzle, or Lego sections. The ends can be pasted together, re-creating a complete double-stranded structure, using an enzyme called DNA ligase. Like restriction endonucleases, DNA ligase does not discriminate. The enzyme pastes together DNA molecules that have been cut by the same restriction enzyme, regardless of their source, or DNA that has been cut by different restriction enzymes that create the same compatible or "sticky" ends. So, using DNA ligase, a human gene fragment can be grafted onto a plasmid, a rat gene can be inserted into a bacteriophage–indeed, any gene combination imaginable can be created, as long as there is a rational scientific or medical objective. Restriction enzymes that cut symmetrically across double-stranded DNA, that do not create single stranded "sticky," ends can also be ligated to each other, but not as efficiently.

These simple steps—restriction digestion and ligation of disparate pieces of DNA—form the foundation of recombinant DNA technology. However, to make it workable, a means for propagating recombinant genes was needed. This was accomplished using plasmids and bacteriophages.

One of the scientific story lines that resulted in the discovery of bacterial plasmids can be traced back to 1902, with observations made by an English physician, Archibald Garrod (1857-1936), who was studying an obscure autosomal recessive genetic disorder of metabolism called alkaptonuria. Metabolic disorders are the "blocked drains" of genetic diseases. The synthesis of vitamins, amino acids, and other key substrates from simple precursor molecules, and the breakdown of spent biochemicals, requires a series of steps, consisting of chemical modifications, before the active substrate or final breakdown product is generated. Each step is guided by an enzyme encoded by a gene, with a different gene coding for each enzyme. Enzymes speed up chemical reactions in cells. Without them, reactions that normally take milliseconds and seconds would take hours or days to complete; the creation of life would have been impossible. Gene mutations that disable an encoded enzyme's function will block metabolic progression. The product on which the enzyme acts will arrest in biochemical limbo and accumulate to very high levels; the product formed by the action of the enzyme will be produced in insufficient quantities. Disease occurs either because the synthesis or breakdown of a critical amino acid or lipid is crippled. During the middle decades of the twentieth century, biochemists were preoccupied with analyzing metabolic pathways in microbes and in man. Today, medical students grudgingly memorize these pathways, almost as a rite of passage, like dissecting a human body, to getting a medical degree.

In alkaptonuria, the genetic defect leads to an abnormality in the metabolic breakdown of homogentisic acid, a dark pigment that builds up and damages different tissues, including skin, bones, kidneys, and the aortic valve. The disease is diagnosed in infants after a few diaper changes because their urine turns black when exposed to air. Garrod observed that the disease was often found in siblings, and yet the parents remained unaffected (the characteristic family profile of an autosomal recessive disorder). At the turn of the 20th century, Thomas Hunt Morgan (1866-1945) and his students at Columbia University were rediscovering Mendel's laws of inheritance using the fruit fly *Drosophila melanogaster* as an experimental tool for the new field of experimental genetics. Morgan discovered that chromosomes carry inherited traits, and he conceived of the existence of discrete genes distributed along the chromosome, like beads on a string. Garrod reasoned that alkaptonuria was due to an inherited gene disorder that led directly to an enzyme abnormality, and that other metabolic disorders were due to similar genetic alterations. It was the first time anyone had ever linked a biochemical process to a gene. But Garrod was a product of his generation, and it was a generation that only understood genes as a concept–the hypothetical transfer unit of inherited traits. No one knew anything about the chemical basis of heredity or how biological information was transferred. Garrod's gene-enzyme connection remained unexplored for nearly four decades.

Garrod's hypothesis was resurrected by a Nebraskan farm boy turned geneticist, George W. Beadle (1903-1989), who trained under Morgan, and a biochemist named Edward L. Tatum (1909-1975), who was an expert at isolating rare metabolites in microbes. The two teamed together in 1941 at Stanford University to produce a classic series of experiments that provided clear proof of the connection between genes and enzymes. Beadle began his career studying seemingly arcane matters; the various colors of Indian corn and the eye color of *Drosophila melanogaster*, the latter a legacy of Morgan's very first genetic insight. Morgan started out as a skeptic of Mendel's ideas of inheritance, believing instead in Lamarck's theory of acquired characteristics, which posited that certain traits could arise in a member of a species due to its environmental circumstances and then be passed down to later generations. Morgan spent two years, from 1908 to 1910, in a failed Lamarckian attempt to develop strains of blind fruit flies by keeping them in total darkness. The endeavor was not a complete waste of time, though: a single mutant fly was found with white eyes instead of red, and transfer of the white eye trait to offspring occurred according to the laws of probability that followed the ideas of inheritance proposed by Mendel decades earlier. This discovery convinced Morgan to give up Lamarck's ideas in favor of Mendel's, and started him on a legendary career that helped launch 20th-century genetics.

Beadle had teamed up with Boris Ephrussi, a Russian-French embryologist, to decipher the mechanism controlling eye color in *Drosophila melanogaster* using various mutant flies that had been isolated by Morgan's "Fly Group" over the years. Through a series of transplantation experiments using imaginal disks, the structural precursor of eyes, Beadle postulated that several enzyme steps were involved in eye color and that genes guided their synthesis. However, the biochemistry of eye color was too complicated to dissect at each step, and they were scooped in the discovery of a key metabolite in the color pathway. After five years of work, the fruit fly was abandoned as a model system for understanding how genes and enzymes are connected.

One day, while listening to a talk Tatum was giving to students on metabolic pathways in microorganisms, Beadle came to the realization he could study the gene/enzyme connection in simple organisms using metabolic and nutritional pathways that had already been well-characterized by biochemists. He and Tatum conceived a simple experimental design using, like the scientists working on penicillin, a bread mold. In this case, the mold was *Neurospora crassa*, an organism that reproduces sexually from spores, whose biochemistry had already been extensively worked out. *N. crassa* is very easy to cultivate in the laboratory–simply mix some mold with a solution of amino acids, sugars, and the vitamin biotin, and the organism does the rest, synthesizing every protein, carbohydrate, and lipid it needs from the elementary building-block precursors provided in the growth medium. Their plan was to create mutant strains of *N. crassa* that had additional nutritional needs, identify the biochemical defect, and cross different strains to work out the genetics. To create nutritional mutants, Beadle and Tatum exposed mold spores to X-rays, which generate new heritable traits by a mechanism unknown to scientists in the 1940s, a method discovered by another one of Morgan's students, Hermann Joseph Muller (1890-1967). We now know that X-rays introduce changes in the genetic code by mutating DNA.

Like a disciplined gambler visiting a Las Vegas casino with only a fixed amount of cash that he or she is willing to lose, Beadle and Tatum gave themselves 5,000 attempts at identifying a nutritional mutant before they would give up. Their plan was to identify mutants by their fussier feeding requirements; instead of being able to grow in the most bare, biotin-supplemented medium, they reasoned, some of the X-ray damaged *Neurospora crassa* spores would require the addition of a vitamin or amino acid that could no longer be synthesized on its own from scratch (auxotroph is the term used for an organism with special nutritional requirements). They hit the jackpot with the 299th irradiated spore-it produced an auxotrophic mold strain that required the addition of pyridoxine, a B-vitamin. Within a few months they isolated other auxotrophic mutants, the most important of which required the amino acid arginine in order to grow. Arginine is synthesized from glucose through a multi-step biosynthetic pathway. At each step along the way, a different organic molecule is generated-ornithine and citrulline are the immediate precursors to arginine. Beadle and Tatum found that some of the *N. crassa* auxotrophs could synthesize arginine and grow when provided with ornithine, some by citrulline, and finally, some required arginine itself. It all depended on which step along the metabolic pathway was deficient (i.e., which gene coding for the enzyme that catalyzes a particular step in the pathway has mutated). By crossing mutant strains, they determined that the nutritional defects they created were due to abnormalities in individual *Neurospora* genes, their now-famous "one gene/one enzyme" hypothesis. This was the first experimental proof demonstrating how genes worked: they were discrete heritable units that made enzymes. Actually, genes direct the synthesis of all proteins, not just enzymes. Beadle and Tatum's dictum should really be "one gene/one protein" or, better yet, "one gene/one polypeptide," since some complex proteins, such as hemoglobin, consist of two protein (polypeptide) chains encoded by different genes. Nevertheless, their partial understanding of gene function was more than sufficient for the time. Along with the experiments of Avery, MacCloud, and McCarty (see chapter 1) demonstrating that DNA transmitted virulence traits in bacteria, the structural framework for how DNA worked began to take shape. These findings, along with the Chase-Hershey blender experiment (see below in this chapter), and the powerful boost provided by the 1953 publication of the double-helix model of DNA proposed by Watson and Crick, launched the age of the gene. It also directly led to the discovery of plasmids by Joshua Lederberg a few years later.

Beadle and Tatum's experiment made a strong impression on Lederberg (b. 1925), a young Columbia University medical student at the time. After receiving his B.A. from Columbia University in 1944, he served a tour of duty with the U.S. Naval Reserve, working in a parasitology lab. There he observed the complicated life cycle and sexual phase of *Plasmodium vivax*, and began to wonder for the first time whether bacteria could also exchange DNA by sexual transfer. The idea was alien to most bacteriologists, who viewed bacteria as primitive creatures that might not even follow the same rules of life as higher organisms, let alone exchange DNA.

Lederberg decided to tackle the problem using a strategy similar to the Beadle and Tatum experiment, and was offered a position in Tatum's new lab at Yale to pursue the idea in 1946-1947. He chose to study the K-12 strain of *E. coli*, which was first isolated from human feces in 1922. Lederberg's plan was to create nutritionally deficient, auxotrophic strains of K-12 and determine if they could be rescued by intact metabolic genes delivered from a normal strain by sexual transfer. Successful rescue would be determined by growing deficient bacteria on agar plates containing a minimal growth medium (i.e., one that was missing the key ingredient); deficient mutants unable to synthesize the key nutrient would die in minimal medium, while rescued bacteria would survive. However, in preliminary experiments, Lederberg found that some deficient bacteria would spontaneously revert back to normal on their own, without the help of normal bacterial genes provided by another bacterium. This self-correction (reversion) can occur, we now know, when a mutant gene mutates back to normal, a rare but measurable occurrence. He had to find some way of distinguishing between his sought-after experimental result (the transfer of genetic information by sexual exchange) and a false positive finding (spontaneous reversion).

Lederberg came up with an experiment using a simple Darwinian survival strategy based on the law of probability. He reasoned that if a double nutritional mutant could be isolated, then the spontaneous reversion of both, involving the mutating of two broken genes back to normal, would occur so infrequently that it would never be observed on its own under his experimental conditions. It would be as likely as getting the winning ticket for a jackpot lottery twice in a row. Lederberg was able to isolate two K-12 strains that had multiple nutritional requirements. One strain was unable to grow unless biotin and the amino acid methionine were added to the growth medium, presumably because the genes responsible for their metabolic synthesis were damaged. The other strain was unable to grow unless the amino acids threonine and leucine and the vitamin thiamine were added. Lederberg then mixed the two strains together, to encourage their coupling, and grew them on agar plates containing growth medium that was deficient in all five nutrients, which would effectively kill off both parent strains. The only way that a bacterium would survive the Darwinian challenge, aside from infinitesimally rare multiple spontaneous reversion mutations, would be through the transfer of intact genes from one strain to another. With large-scale transfer of genetic information through bacterial sex, the intact genes in each deficient strain should complement the other, like two jigsaw puzzles missing different pieces that would form one complete puzzle if combined.

The experiment worked perfectly. After mixing together the two *E. coli* strains, Lederberg was able to easily detect hundreds of colonies after overnight growth on a minimal medium. His conclusion: the two imperfect strains had exchanged their intact metabolic genes to create new bacterial variants. It was the only way to explain the findings. Lederberg had uncovered a key microbial survival strategy—gene transfer between bacteria—enabling them to thrive and diversify, and, as microbiologists would later learn, to plague humans by transferring virulence factors and antibiotic resistence genes.[34]

---

[34]In 1958, at the age of 33, Lederberg was rewarded for this discovery, sharing a Nobel Prize for Physiology or Medicine with Beadle and Tatum.

What was the mechanism behind genetic exchange in bacteria? It was a tangible, measurable process, but it was invisible (it would take another 15 years of research before plasmids could be isolated and visualized using electron microscopes). However, even though Lederberg could not see what was happening, he was able to determine that the transfer of heritable traits in K-12 required physical contact between bacteria. He also showed that the transfer of information behaved like an infection: not only did a low number of "male" bacterial donors placed in a culture of "female" recipients result in the transfer of heritable traits, but females were rapidly and efficiently converted into males. The transmissible agent, which Lederberg called the F-factor (fertility factor), contained genes that corrected nutritional deficiencies and also provided the information for self-assembly. We now know that the F-factor is a conjugative plasmid, a moveable piece of DNA bacteria can use to transfer genetic information to other bacteria.

A similar type of invisible genetic delivery system was uncovered in 1961 when Japanese researchers Tsutomu Watanabe and Toshio Fukasawa found that a strain of *Shigella* recovered from a patient whose infection was not cured with chloramphenicol, an antibiotic, could transmit the resistance trait to non-resistant strains, analogous to F-factors. They called the heritable unit an R-factor (resistance factor). It was the first antibiotic resistance plasmid. After R-factors were discovered, bacteria that had been frozen in the pre-antibiotic era were thawed out, and many of them were found to carry antibiotic resistance plasmids. Resistance plasmids, it turns out, have been around for eons. Bacteria had been using them as a way of protecting themselves against the antibiotics produced in nature by other microorganisms, well before humans arrived on the scene; we merely accelerated their expansion by natural selection when antibiotics gained worldwide use in treating infectious diseases in humans and animals.

The idea that plasmids could be used to create and propagate novel gene combinations was the brainchild of Herbert Boyer and Stanley Cohen, scientists at Stanford University and the University of California at San Francisco, respectively. Boyer was working on restriction endonucleases when he heard a presentation on plasmids by Cohen at a meeting in Hawaii in 1972 and realized the potential of fusing their two areas of interest, literally and figuratively. Working with a plasmid containing a gene for ampicillin resistance, Boyer and Cohen inserted a gene from frogs using DNA ligase. Both the plasmid and frog gene had been cut with the restriction enzyme *EcoR1* to create compatible ends suitable for pasting with DNA ligase. The plasmid was then introduced into a harmless strain of *E. coli,* which was grown on agar plates containing ampicillin. Only bacteria that accepted the plasmid with the antibiotic-resistance gene survived, along with the inserted foreign gene (which was only along for the ride). The parent strain without plasmid was wiped out by the ampicillin. Resistant bacterial colonies or clones were easily isolated, and then grown to obtain large quantities of plasmid DNA, together with the inserted foreign gene. For the first time, humans had intentionally cultivated and harvested a gene, a process referred to as recombinant DNA, molecular cloning, or just cloning.[35] Although the growth of the first recombinant gene using a plasmid was of no immediate practical importance, it was a dramatic "proof of principle" experiment that was sufficient to launch the new technology in the scientific world.

Although Boyer and Cohen initiated molecular cloning in plasmids, they were not the first to create recombinant genes. That distinction belongs to Paul Berg, who, a year earlier, fused together DNA from bacteriophage and SV40, a commonly used virus that infects monkey and human cells, and causes cancer in laboratory animals. The basic strategy of using enzymes to piece together DNA from different organisms began with his studies. Berg stopped short of growing recombinant SV40 genes in bacteria because he was concerned about the possibility of accidentally creating pathogenic viruses through his manipulations. Later, he proposed a moratorium on all recombinant DNA research until safety concerns could be addressed. Eventually he helped develop the NIH safety guidelines that govern recombinant

---

[35]The term "cloning" was initially meant to describe the isolation of genes from individual "clones" of antibiotic-resistant bacteria. The word has taken on a completely new meaning today. It is more commonly interpreted in the context of "cloning" animals using nuclei from living cells implanted into ova to create a duplicate copy of the nucleus donor.

DNA research to this day. Berg went on to share a Nobel Prize, a distinction denied to Boyer and Cohen. The Nobel committee probably did not approve of the direction their research was taking: Boyer and Cohen were aware of the commercial implications of their invention and, with the help of venture capitalists, started the first biotech company, Genentec, to develop medical products from recombinant DNA. Instead of the glory of a Nobel Prize, Cohen and Boyer had to be content with the accolades of stockbrokers and investors, and the patients and scientists who benefitted from their entrepreneurship; they are acknowledged as the founding fathers of the biotechnology industry.

A few years after plasmid-mediated recombinant DNA was invented, several groups improved this technology by developing bacteriophage as a gene-cloning system. Bacteriophages have the advantage of being able to incorporate larger fragments of DNA than plasmids. This provided researchers with the ability to create entire "libraries" housing the genomes of complex organisms, such as mammals, in a single recombinant DNA experiment.

Bacteriophages were first detected in 1896, when a British chemist, E. H. Hankin, noticed that water from the Ganges and Jumma rivers in India contained a substance that could kill *Vibrio cholerae* grown in culture. This turned out to be a *Vibrio cholerae*-specific bacteriophage, a finding that was not appreciated at the time; it took another 20 years before bacteriophages were cultivated and subjected to more intense scientific scrutiny, with the discoveries of bacteria-killing extracts by Frederick W. Twort in 1915 and Félix D'Hérelle in 1917. Twort was interested in "essential substances," mysterious factors he believed were needed for the growth and sustenance of pathogenic organisms. He was also interested in developing a culture system for growing the *Vaccinia* virus used in the smallpox vaccine. At the time, *Vaccinia* was cultured from the skin of cows, which sometimes contaminated the vaccine with *Staphylococcus*. He believed that bacterial extracts might contain the essential substance needed for *Vaccinia* growth that could be included in a recipe for growing the virus in culture, as a way of avoiding the problem of contamination. His experiments failed. However, in the course of his work he noticed that agar plates covered with a carpet of bacteria had developed bacteria-free holes–punched-out, pupil-size voids. These were found to contain an invisible bacteria-killing infectious agent (bacteriophage) that could be cultivated. Twort recognized the potential of his discovery and spent the rest of his 35 years of life trying to establish a therapeutic use for it. Bacteriophages were discovered independently in 1917 by D'Hérelle, who was also intrigued by the therapeutic possibilities. D'Hérelle believed that bacteriophages were produced as part of the body's natural defense against infectious bacteria. He traveled around the world searching for novel bacteriophage in the feces of people who had overcome deadly intestinal infections. A bacteriophage that infects *Vibrio cholerae* was actually used to treat cholera outbreaks in India in the 1930s, and Los Angeles in the 1940s, with some success. In the 1930s, the Eli Lilly drug company developed a bacteriophage impregnated gel as a topical treatment for *Staphylococcus* skin infections. However, inconsistent results in patients and the emergence of easy-to-manufacture and well-tolerated antibiotics made bacteriophage therapy obsolete. By the end of his research career, Twort had become a bitter, irrelevant scientist who was kept on at his London facility, the Brown Animal Sanitary Institution, more because of seniority than scientific productivity. He was eventually turned into a pensioner in 1944 when the Nazis bombed the institution during the Blitzkrieg.

With the emergence of widespread antibiotic resistance, though, bacteriophage therapy has returned as a legitimate scientific pursuit, and it is very likely that in the next few years, it may become a treatment option for patients infected with antibiotic resistant bacterial strains. Some advocates have even proposed using bacteriophage as a safe, natural alternative to antibiotics in ridding the food supply of potentially dangerous pathogens, like *Salmonella* strains that cause food poisoning. Imagine, if you can, a *Salmonella*-destroying bacteriophage added to poultry meat destined for a chicken hot dog factory. Can bacteriophage-laced mouthwash that destroys the bacteria responsible for bad breath be far off?

During the late 1930s and 40s, a number of scientists, primarily under the influence of Max Delbrück (1906-1981), began to study bacteriophage as a tool for understanding viral function and genetic information transfer. Delbrück was a German theoretical physicist who trained under Neils Bohr. He was an early member of an expanding club of physicists, later joined by Francis Crick and others, who changed career direction to become world-class biologists (whereas the opposite career move, from biologist to world-class physicist, almost never occurs). Delbrück first became interested in biology as a result of discussions he had with Bohr, who thought that some of the quantum mechanics principles he had proposed would apply to biological systems. Delbrück's interest in bacteriophage genetics was stimulated by informal discussions that took place in the mid-1930s at the Kaiser Wilhelm Institute for Chemistry between physicists and biologists who were bored with the official Nazi-sanctioned seminars being offered. In 1937, before the outbreak of World War II, Delbrück left Germany and moved to the United States, eventually becoming a citizen there. He developed a simple culture system for growing bacteriophage, and in the early 1940s, Delbrück (while teaching at Vanderbilt University) teamed with Salvador Luria (who was at the University of Indiana) to discover that new types of viruses can develop when genes from different viruses recombine.

Bacteriophage genetics played an important role in providing the final experimental proof that the transfer of information that controlled the inheritance of traits and physical processes resided in DNA. To prove this, Alfred Day Hershey and Martha Chase came up with their exquisitely simple "blender experiment." Bacteriophages are simple organisms constructed from a few proteins surrounding a single stretch of DNA. They come in a variety of whimsical shapes resembling lollipops, mosquitoes, and lunar landers. Their protein coats attach to bacteria by binding to specific receptors, after which genetic material is injected into the host bacterium. During the 1940s and early '50s, it was not known whether viral proteins or viral DNA entered bacteria to orchestrate genetic information transfer. The experiments of Avery, MacCloud, and McCarty and those of Beadle and Tatum put DNA at the top of the list, but the idea was not accepted by every scientist, and experimental proof in bacteriophage was lacking. Chase and Hershey used a bacteriophage strain named T2 to test the opposing hypotheses. They divided T2 into two groups. One was tagged with a small amount of radioactive sulfur using the amino acid cysteine, which labels the bacteriophage's proteins. The other was tagged with nucleotides containing radioactive phosphorus, which labels DNA. Both groups were added to solutions of bacteria for a sufficient period of time to cause infection and genetic transfer. The bacteria were then agitated in a milkshake blender; viruses adhering to bacteria can be jarred loose with physical force, thus interrupting genetic transfer. If viral proteins transferred genetic information, radioactive sulfur would be found inside the bacteria; whereas if DNA was responsible, radioactive phosphorus would be found. The experiment worked brilliantly: only radioactive phosphorous was detected inside the transformed bacteria, while radioactive sulfur could only be found in the bacteria-free wash solutions. It was the final experimental proof needed to firmly establish that DNA carried genetic information.[36]

It turned out that recombinant DNA principles could also be applied to bacteriophage. Its DNA could be isolated and cut with restriction enzymes, and reconfigured with non-viral genes using DNA ligase. However, bacteriophage DNA, in its native state or as a recombinant molecule containing foreign pieces of DNA, is unable to efficiently infect bacteria on its own. In order for recombinant bacteriophage to propagate properly, the entire virus–that is, its DNA and the DNA's accompanying proteins– would have to be reconstructed. In 1975, Andrew Becker and Marvin Gold invented a method for completely reconstituting live bacteriophage from recombinant bacteriophage DNA and viral proteins, to create infectious particles that could be used to efficiently transfer foreign DNA into bacteria–a milestone finding. Essentially, they provided a recipe for assembling a living virus from its nucleic acid and protein components, duplicating a process that viruses have been able to implement on their own, via infected host cells, working the details out over hundreds of millions of years of evolution. With the test-tube assembly of

---

[36]In 1969, the Nobel Prize in Physiology or Medicine was awarded jointly to Delbrück, Luria, and Hershey for their discoveries concerning "the replication mechanism and the genetic structure of viruses."

bacteriophage, a new vehicle for cultivating useful genes in bacteria was developed, extending the range of recombinant DNA.

Today, scientists can shop for plasmid and bacteriophage products best suited for their experimental needs, thumbing through biotechnology company catalogues and scrolling through Web sites: plasmids and bacteriophages with a built-in color-coding feature for easy screening; plasmids with the capacity to express RNA and protein in cell-free systems mimicking the function of ribosomes; bacteriophage that can express, in bacteria, the mammalian proteins encoded by recombinant genes; and hundreds of others. There is even a family of plasmids containing the gene encoding the enzyme that makes fireflies glow (firefly luciferase). Gene promoters inserted in front of the luciferase gene can activate its expression and cause light to be emitted; the more powerful the gene promoter, the more light is emitted. The luciferase system is used by scientists who study how genes are regulated.

Today, tens of thousands of scientists around the world are using some application of recombinant DNA for their research. Manipulating genes for a scientific end, using tools borrowed and stolen from microbes, has become almost as routine a part of biology research as turning on a Bunsen burner.

CHAPTER 15

Reversing the Code

Another key ingredient in the recombinant DNA cookbook derived from the infectious disease world is reverse transcriptase, an enzyme that converts RNA into DNA. All the genetic manipulations used to create and cultivate recombinant molecules, the cutting and pasting of genes and gene fragments, require DNA; but RNA–specifically messenger RNA (mRNA)–is needed for one of the most important applications. Messenger RNA is crammed with genetic information, including all the relevant coding information needed to make proteins (see chapter 1). This is especially true for large, complex genes such as dystrophin, a damaged version of which causes muscular dystrophy. The entire dystrophin gene, the largest in the genome, is divided into 79 exons and introns spread out over a span of two and a half million nucleotides (genes typically have fewer than a dozen exons and introns, spanning approximately 50,000 to 100,000 nucleotides). Yet dystrophin mRNA, which is formed when intron RNA is spliced out, contains only fourteen thousand nucleotides, less than 1 percent of the gene. For scientists interested in dystrophin protein–the only part of the gene that could possibly be used to treat muscular dystrophy someday–working with the entire gene would be like having to sit through hundreds of hours of unedited film footage to view the two hours that eventually form the final, edited cut. Moreover, the coding regions, from which mRNA is derived, are the gene segments most commonly damaged in genetic disorders. However, in order to insert mRNA into a plasmid or bacteriophage, which scientists would to do in order to plasmid or bacteriophage which scientists would like to do in order to propogate the protein-coding portion, it has to be converted into a DNA copy first, since RNA is unable to directly insert itself into DNA genomes. Reverse transcriptase provided this option. The enzyme converts RNA into a precise DNA replica, called complementary DNA (cDNA), which can be engineered into plasmids and bacteriophage almost as easily as a restriction endonuclease gene fragment.

In catalyzing the conversion of mRNA into cDNA, reverse transcriptase reverses the natural flow of genetic information, which is from DNA to RNA to protein–a pattern so constant in life on earth it is referred to as the "central dogma." However, this notion of information flow applies only to DNA life forms. For retroviruses, which have RNA genomes, the transfer of genetic information is reversed: retroviruses convert their RNA life code into a DNA copy intermediate in order to both replicate and integrate into a host genome. It stands to reason that these viruses should have a mechanism in place to accomplish this conversion–a reverse transcription process. Indeed, retroviruses are nature's only natural source of reverse transcriptase. But the ability to harvest the enzyme came 60 years after scientists first encountered a retrovirus. It was worth the wait. The discovery produced a scientific windfall: reverse transcriptase not only expanded the range of recombinant DNA technology, providing a host of important therapeutic biological products by enabling scientists to incorporate cDNAs into plasmids and bacteriophage, but also played a key role in identifying the genes that cause cancer and in discovering HIV.

The remarkable story of reverse transcriptase began quite innocently in 1911 when a New Jersey farmer, who was having a problem with lumpy chickens, paid a visit to Peyton Rous (1879-1970), a young physician and researcher at the Rockefeller Institute for Medical Research (now Rockefeller University), in New York City. The lumps were sarcomas–cancers involving the soft tissues of the body, such as muscle and connective tissue– and they were spreading among the chickens like a contagious infectious disease. Rous had graduated from the Johns Hopkins Medical School in 1905 and came to the Rockefeller Institute in 1907. He was a trained pathologist and microbiologist who was persuaded by Simon Flexner, a renowned scientist and director of the institute, to pursue cancer research.

Rous was able to show that a transmissible agent smaller than bacteria caused chicken sarcoma: injecting a cell- and bacteria-free filtered extract of tumor tissue into healthy chickens rapidly induced the tumors. Although his experimental findings were indisputable, most scientists–and even Rous himself in the beginning–questioned the idea that an invisible infectious agent caused chicken sarcoma. The prevailing view was that cancer was a cellular problem and not an infectious disease. It was argued that the cancer induced by Rous's cell-free extract was caused by a toxic cellular product released from dead cancer cells. Even after the virus was isolated some decades later–and christened the Rous Sarcoma Virus (RSV)–the idea that viruses could cause cancer was met with great skepticism. Eventually though, as other examples were discovered, it became clear that some viruses were indeed masters at transforming normal cells into cancerous growths, at least in lower animals. In one of the more dramatic examples, it was shown that female mice with mammary-gland tumors were able to transmit the disease to their suckling pups by passing a virus called MMTV (mouse mammary tumor virus) through their milk.

Despite the findings in rodents and chickens, doubts remained about a larger role for viruses in human cancer. There was no evidence for a transmissible infectious agent, either experimentally or in epidemiological studies, with the exception of a rare cancer found in Africa called Burkitt's lymphoma, which is caused by a variant strain of the DNA virus responsible for infectious mononucleosis, and liver cancer associated with viral hepatitis.

Discouraged by the negative reaction to his findings, Rous stopped pursuing cancer research in 1915, instead turning to other problems in physiological pathology. During that time, he made major contributions in establishing a cell-culture system for mammalian cells, which became an instrumental experimental tool for cancer researchers years later, and had the idea for developing blood banks, repositories for storing blood that could be used, on an as-needed basis, for transfusions.[37] Before World War I, transfusions were administered directly from donor to

---

[37]Rous was at Rockefeller when Karl Landsteiner was there, doing research on blood groups (see chapter 5).

recipient because there was no effective method for storing blood. Along with Joseph Turner, Rous tested various anticoagulants and cooling systems, and found a way to store blood safely for a month at a time. Turner used this knowledge to establish the first blood bank, behind enemy lines in France during World War I, where the standard method of direct person-to-person transfusion was impractical. Rous, ever curious, also spent much of his research time on matters more mundane than cancer and blood banks: he was interested in gall bladder function and the formation of gallstones, and made important contributions in these areas.

When he returned to cancer research in 1934, Rous spent many years trying to identify cancer-causing retroviruses in humans, but was unsuccessful. It turned out that Rous's detractors were correct in their assessment that the causes of human cancer are almost always cellular genetic problems, not retroviruses (with a few notable exceptions). Ironically though, as described below, RSV still proved to be one of the keys to revealing what causes cancer in humans. Because of the importance of his discovery, Rous was awarded a Nobel Prize in Physiology or Medicine (which he shared with Charles B. Huggins) in 1966, at the age of 87, more than half a century after his first encounter with the New Jersey chicken farmer; he was one of the oldest recipients of the award.

Retroviruses were all the rage in the scientific world in the 1960s and 70's. All viruses are intracellular parasites, incapable of performing any metabolic or physiologic function. The only thing they do is invade cells and deliver their RNA genetic package; the infected cell does the rest, providing all the nucleotide and amino-acid building blocks for synthesizing the genes and proteins needed to fabricate new viral particles. Of all the viruses found in nature, retroviruses win the prize for genomic parsimony. They are the simplest forms of life (except perhaps for viroids, which are infectious RNA particles devoid of protein that cause disease in the plant world, and are arguably not life forms at all). The most common types of retroviruses have only three genes. Even the more complicated ones, such as HIV, barely have a dozen. For all the elegant simplicity of retroviruses, it was the mechanism they used to convert their RNA genome into DNA, a necessary step in their replication, that intrigued scientists most; the enzyme that reversed the usual direction of how genetic information is transferred had no counterpart in the bacterial and eurkaryotic worlds. How was this reversal of the central dogma accomplished?

Two young researchers, David Baltimore (b. 1938) and Howard Temin (1934-1994), both former students of virologist Renato Dulbecco, solved the puzzle when they independently isolated reverse transcriptase, using standard biochemical purification techniques. It proved to be a fairly straightforward process. The discovery was there for anyone to take, but it was Temin and Baltimore who had the insight and drive to get there first. When hindsight reveals that a profoundly important discovery was achieved fairly easily, using simple scientific reasoning and experimentation, there is a collective slap to the heads of thousands of scientists chastising themselves for not doing the obvious. Many of us were slapping ourselves when the PCR technique was invented more than twenty years ago; it was so simple and elegant, its influence on all aspects of genetics so profound, from research to diagnostics and forensics. It was particularly irksome, though, that the genius behind the discovery, Kary Mullis (b. 1944), was an eccentric maverick who left science after winning a Nobel Prize in 1993 to write, surf, rollerblade, and proselytize that HIV did not cause AIDS.

The papers describing Temin's and Baltimore's discovery appeared in a single issue of the scientific journal *Nature* in 1970, one following the other. Because of Baltimore's political activism, though, he came close to forfeiting his share of the bragging rights to what turned out to be one of the more important discoveries in biology–one that would earn him and Temin a Nobel Prize in just five years, catapulting Baltimore to the very top of the scientific pecking order. Baltimore isolated reverse transcriptase in 1970, and was just finishing his experiments when President Richard Nixon ordered American troops to invade Cambodia, at which point Baltimore halted his work for a week to take part in campus-wide demonstrations at MIT, where he was an associate professor. After returning to the lab, he successfully completed his experiments and quickly sent off a manuscript to *Nature*. It arrived just in time, only days ahead of a paper submitted by Temin, who, independently of Baltimore, had also isolated reverse transcriptase. *Nature* published the papers in the same issue, which it often does when two or more manuscripts covering the same topic arrive at their editorial offices at nearly the same time. If Baltimore's final experiments had not gone smoothly and additional laboratory time had been needed to confirm his results, the time he took off from the lab to protest could have been costly for his career. Being the first to discover reverse transcriptase and the rewards that came with it might have belonged to Temin alone.

Both Temin and Baltimore went on to make important contributions in virology, immunology, and AIDS research. Baltimore later became the director of three of the most prestigious scientific establishments in the United States: MIT's Whitehead Institute, Rockefeller University, and, most recently, Cal Tech. However, he became embroiled in a scientific fraud case during his Rockefeller tenure in the early 1990s, involving the author of a paper published in the journal *Cell* in 1986 on which he was a co-author. The paper, which dealt with a novel finding that involved the effect that foreign genes have on immune responses when introduced into mouse embryos, was based primarily on the work of a collaborator, Thereza Imanishi-Kari (a Japanese researcher from Brazil working at MIT at the time, who later moved to Tufts University). Doubt was cast on the results when the experiments could not be successfully reproduced by Margot O'Toole, a postdoctoral fellow working in Imanishi-Kari's lab. O'Toole believed that the negative results were due to errors in the data originally presented in the *Cell* paper, and she claimed that lab notebooks actually contained data that contradicted the reported results. But, Imanishi-Kari and independent reviewers from MIT denied her request that the *Cell* paper be retracted; although there was an error in labeling one of the figures, this was viewed as too minor to affect the main idea. Baltimore agreed.

The relationship between a senior mentor and junior postdoctoral fellow can occasionally be contentious, especially when there is a clash of strong-willed personalities. Boiling points are reached when a postdoctoral fellow makes important discoveries with little input from his or her mentor but the credit is lavished entirely on the more senior person without due acknowledgment, or when there is suspicion of scientific fraud on either side. O'Toole and Imanishi-Kari had problems from the very beginning of their relationship. Imanishi-Kari felt that the experiments were not successful because O'Toole was not working hard enough, although, as a new mother, O'Toole had a legitimate excuse for not putting in the grinding hours that most postdocs devote to their work. O'Toole, however, suspected that she was unable to replicate the original results because of the experiments' incorrect conclusions based on scientific fraud. No stranger to controversy, O'Toole had been dismissed from her previous postdoctoral position and had had a series of confrontations with administrators and lab co-workers. Soon after her arrival in Boston, she became involved as a witness to an alleged police brutality case against the Boston Police force. Her problems with Imanishi-Kari escalated and remained unresolved. O'Toole felt that she had no other recourse except to become a whistle-blower.

Eventually, the case was brought to the attention of the chairman of the House Energy and Commerce Committee that oversees the NIH, Congressman John Dingell. Dingell had a reputation for doggedly pursuing cases of scientific fraud by investigators who used federal funds for their work. A congressional hearing followed, at which Baltimore testified. Baltimore could not hide his irritation at the committee and at the government's meddling in what amounted to an issue of scientific accuracy that only scientists could judge, and was perceived by the committee as an arrogant elitist. Dingell felt publicly humiliated by Baltimore and decided to crush him at any cost, even to the point of using the forces of the U.S. government usually reserved for murderers, counterfeiters, spies, and terrorists. Dingell ordered the Secret Service and the F.B.I. to investigate Imanishi-Kari's lab notebooks, to dig out the scientific fraud he was convinced was hidden in its pages. Walter Stewart and Ned Feder, two scientist-watchdogs working in the NIH Office of Scientific Integrity, and consultants to Dingell, who believed that fraud was rampant in the scientific community, also got involved in the investigation. Helping to bring down a scientist of Baltimore's stature would provide a test case for their views.

Baltimore found himself in a no-win situation and was urged by prominent scientists, including James Watson, to drop his support for Imanishi-Kari. An interim report by the NIH in 1994 suggested that the data reported in the *Cell* paper were fabricated by Imanishi-Kari. However, this was based on an incomplete analysis of her laboratory notebooks, and she was found innocent two years later, following an appeal. The second decision was highly critical of the original NIH verdict. After an investigation costing tens of millions of dollars and taking 10 years, Imanishi-Kari was finally cleared of charges. She was reinstated at Tufts, with her NIH grants restored. But she had lost 10 years of work during what should have been her most productive period. O'Toole, marked as a troublemaker, was unable to return to an academic scientific life, since no one would hire her. As for Baltimore, his steadfast defense of Imanishi-Kari and prominent role in the most public examination of scientific fraud in history proved too much for the Rockefeller Board of Directors, who, like most university and foundation leaders, hate controversy. Baltimore was forced to resign. Fortunately, his scientific and administrative career did not suffer long-term damage: he has been the president of Cal Tech since 1997, and he has continued to make important contributions in immunology and the search for an AIDS vaccine.

The discovery of reverse transcriptase had an immediate impact on three emerging areas of research in the 1980s: recombinant DNA, the role of retroviruses in human disease, and the genetics of cancer.

The incorporation of reverse transcriptase into recombinant DNA methodology allowed scientists to grow the coding portion of genes in bacteria or yeast and to harvest therapeutic quantities of useful proteins that could not be easily isolated or synthesized. One of the first was human insulin. For the past 15 years, most diabetics requiring insulin have been controlling their elevated blood sugar by injecting themselves every day with human insulin synthesized through recombinant DNA technology, instead of the cow and pig insulin that had been used since the 1920s, extracted from the pancreases of millions of slaughtered animals. Human insulin works better on human cells, enhancing control of blood sugar; moreover, diabetics sometimes developed antibodies to the animal-derived insulin used in the past, which made it ineffective.

Recombinant DNA based on reverse transcriptase has provided doctors with human growth hormone, which is used to treat short stature, and with erythropoietin, the hormone that boosts production of red blood cells in patients with chronic renal failure and in athletes seeking an illegal competitive edge (see chapter 4). Before the advent of recombinant DNA, growth hormone was extracted and pooled from human pituitary glands. A few patients treated with human pituitary extract came down with CJD (Creutzfeldt-Jacob Disease), the human equivalent of bovine spongiform encephalopathy- "mad cow disease." Pituitary extracts were derived from thousands of autopsy samples, and some were contaminated with prions (proteinaceous infectious particles that lack nucleic acid), responsible for the Swiss cheese-like brain degeneration characteristic of CJD and mad cow disease.

Some recombinant products are used to help fight serious infections, in a sense, turning the table on infectious organisms by exploiting the agents of virulence and antibiotic resistance to create microbe-killing products. One example is the white-blood-cell growth factor GM-CSF, the primary hormone that regulates the growth and development of bacteria-killing phagocytes. The gene coding for GM-CSF was isolated in the 1980s, which, after cloning in bacteria and yeast, allowed scientists to harvest large quantities of the hormone. It is now used in patients whose white blood cell counts have dropped to dangerously low levels due to chemotherapy, radiation treatment, or disease. Pharmaceutical companies have also used recombinant DNA to synthesize large quantities of interferons, antiviral factors normally produced by some types of T-lymphocytes. One variety of interferon is being used to treat the chronic hepatitis C infections of hundreds of thousands of people in the United States. It is also used to treat malignant melanoma and multiple sclerosis.

Recombinant DNA has also provided scientists with a new way to manufacture vaccines. There are several different approaches, one of which is to take genes coding for microbial antigens that do not evoke a strong immune response and insert them into organisms that are powerful immune system activators, such as *Vaccinia*. The immune system is fooled into generating a similarly powerful response against the proteins encoded by the inserted passenger genes. Several experimental HIV vaccines have been developed using this approach.

Another outcome related to the discovery of reverse transcriptase has been the identification of retroviruses as a cause of infectious disease. Viruses, especially retroviruses, are much more difficult to grow in culture than bacteria. Bacteria are capable of growing on their own using simple sources of nutrients. Usually some formulation containing agar, filled with sugars, salts and amino acids is used, even a thin potato slice will do. However, viruses need cells in which to grow, and often require a very distinct cell type. In addition to being difficult to cultivate, retroviruses are also very small, hard to detect even with electron microscopes. Consequently, implicating a retrovirus as the culprit responsible for an emerging infectious disease or a cancer can be quite demanding. This is where reverse transcriptase is helpful. A simple and specific test is available to screen for the presence of reverse transcriptase in the body, which would suggest that a retrovirus is there. In 1983, AIDS was transformed from a mysterious disease, apparently afflicting only a subgroup of gay men and Haitians, to a worldwide retroviral pandemic after Luc Montagnier at the Pasteur Institute in Paris detected reverse-transcriptase enzyme activity in cultured T-lymphocytes taken from a lymph node of a gay man whose immune system was severely impaired. Within a year, Montagnier and Robert Gallo's group at the NIH were able to isolate HIV, and soon thereafter developed the earliest blood tests. It was an important achievement. Transfused blood could now be screened for the presence of HIV, and individuals at high risk could determine if they had acquired the virus. Unfortunately, controversy followed. Gallo became embroiled in a patent dispute with the French concerning the development of the HIV blood test, involving the rights to hundreds of millions of dollars in royalties, which became as publicized, contentious, and drawn-out as the Baltimore affair. Gallo, like Baltimore before him, became the object of an investigation by the investigative trio of Representative Dingel and the NIH's Stewart and Feder. Gallo claimed that the HIV viral strain used to develop the AIDS test was isolated by his lab, a claim supported by several NIH officials, including Bernadine Healy, the NIH's director at the time; but analysis showed that the virus was identical to the one originally sent to his lab by Montagnier. Soon after Montagnier isolated the virus and named it LAV (lymphadenopathy-associated virus), he graciously provided samples to Gallo, one of the world's premier virus researchers, for further analysis. After initially disputing the findings, Gallo finally had to admit that the genetic evidence supported the French claim; the HIV test he had patented had been developed using the viral strain originally sent by Montagnier and the Pasteur Institute. He eventually claimed that the scientists and assistants in his lab who were assigned to the project had accidentally cross-contaminated the lab's native stocks of HIV with the Pasteur Institute virus, and had unwittingly developed the AIDS test with it. Although this was a plausible explanation, considering how common it is to cross-contaminate cell lines if several are being cultivated at once, many people did not believe the story. After a decade of lawsuits and congressional hearings, the royalty agreement was rewritten in favor of the French group; and Gallo was forced to leave the NIH. Two of the crowning achievements of the 1980s–the isolation of the virus responsible for the world's worst and most enduring modern pandemic and the rapid development of a diagnostic test for that virus–were tarnished.

Reverse transcriptase has also had an impact on HIV treatment; AZT, the first effective drug developed against HIV, is the senior member of the class of reverse transcriptase inhibitors that block HIV replication by preventing the conversion of viral RNA to its DNA copy.

The discovery of reverse transcriptase has had a powerful impact on cancer research. Although retroviruses are rarely causes of cancer, understanding their mechanism of action in lower animals led to the discovery of oncogenes, the genes responsible for cancer in humans. Paradoxically, it was the absence of *pol* (the retroviral gene coding for reverse transcriptase) that played an important role in finding the very first oncogene. Fittingly, RSV was the principal player.

RSV, for all its power in being able to transform a normal cell into a cancerous growth, is a weakling when it comes to replicating itself-it can not. This "replication deficiency," as in many other cancer-causing retroviruses, is because its *pol* gene is missing. The absence of *pol* forces these retroviruses to rely on intact retroviral partners–helper viruses–to supply the missing reverse-transcriptase enzyme needed for their propagation. During the 1970s, through the work of Hidesaburo Hanafusa, Michael Bishop, Harold Varmus, and other researchers, molecular analysis revealed that the *pol* gene of RSV and other replication-deficient, cancer-causing retroviruses had been replaced by genes captured from the animal cells they infected. These were named oncogenes, and were found to be essential components in the retroviruses' cancer-causing program. Cutting them out of the virus crippled their ability to rapidly convert a normal cell into a cancer: they could still do it, only much more slowly.

Oncogenes are mutated versions of cellular genes that are normally involved in regulating cell growth and gene expression. When mutated or under the control of retroviruses, oncogenes are expressed at abnormally high levels or encode an abnormal protein that overstimulates DNA replication and cell division. The very first oncogene to be found, *src* (pronounced "sark"), was identified as the cancer-causing gene in RSV. *Src* is not a natural retroviral gene. RSV procured it in some long-ago passage through an infected chicken's genome. Within a year or two of the discovery of *src*, scientists had isolated a half dozen other oncogenes that had been similarly captured by retroviruses. In each instance, the inclusion of an oncogene dramatically increased the speed and efficiency of their cancer-causing capability. These are exciting findings, but what was the connection to human cancers, which are only rarely caused by retroviruses?

The relationship became apparent with a series of rapid-fire experimental findings, one of which was a remarkable observation made by several research teams led by Robert Weinberg, Michael Wigler, and Geoffrey M. Cooper: DNA extracted from some human cancers had the capacity to transform mouse cells that were growing in culture into malignant tumors. DNA can be introduced–"transfected"–into cultured cells derived from many different species, using a variety of different strategies. When DNA from a human bladder carcinoma cell line called T24 was transfected into a commonly used mouse cell line, NIH3T3 fibroblasts, a small population of them formed clusters of malignant-appearing cells. When the clusters were isolated and injected into a mouse strain used by cancer researchers, the "athymic nude mouse," cancers rapidly developed.[38] DNA from normal human cells failed to transform NIH3T3 cells. The DNA from the bladder-cancer cells was behaving as if it contained a cancer-causing infectious agent, but it did not. It was a single mutant gene, and it was identified using a type of Darwinian selection strategy, along with help from human junk DNA.

Years of experience introducing large blocks of genes and whole chromosomes into cultured cells showed large chunks of extraneous, non-essential DNA are lost as cells divide, unless a selective survival element is introduced that forces the retention of a critical foreign gene. For the bladder cancer experiment, the Darwinian selection component was a human experimenter picking out clusters of transformed mouse cells and discarding the rest. Only NIH3T3 cells that retained the unknown cancer-causing gene survived, because they were selected by the careful hand of the experimenter. Clusters were grown and maintained, DNA was extracted, and the transfection experiment was repeated. With each successive step, less and less extraneous human DNA was retained; the essential genetic factor responsible for causing cellular transformation was whittled down from three billion nucleotides (the full complement of human DNA) to a few hundred thousand. The human cancer-causing gene was eventually isolated, using a simple method for purifying a small patch of human DNA from a large pool of mouse DNA, based on the presence of human-specific repetitive sequences in junk DNA. After cloning the gene, using standard recombinant DNA techniques, its DNA sequence was determined and found to be

---

[38]The athymic nude mouse has a double genetic whammy. It is born without a functional thymus gland, which prevents it from developing a normal T-lymphocyte immune system, making it prone to cancer; and it is completely hairless.

exactly the same as the viral oncogene *H-ras*, which had been isolated a year earlier from a mouse retrovirus called the Harvey sarcoma virus. They were the same exact gene, down to the very last nucleotide, including the single base change that distinguished it from its normal *H-ras* counterpart in the human genome, a mutation at codon number 12. This change was ultimately shown to enhance the encoded protein's effect on cell growth. The bladder-cancer cells harbored the defective gene, but the *H-ras* genes in unaffected cells from the same patient were completely normal. A mutation arising by chance in a single bladder-mucosal cell at *H-ras* codon 12 had bestowed the gene with the capacity to transform a normal cell into a deadly cancer.

There was one glitch in the experiment. The NIH3T3 cells used in the experiment are not normal. Normal cells have a limited capacity to divide, and die in culture after a few weeks. This is an inevitable feature of cellular life, which scientists call senescence. However, NIH3T3 cells–and others that are maintained perpetually in culture for research purposes–grow without undergoing senescence. Although NIH3T3 cells do not cause malignant cancers in mice, unless provided with oncogenes, they do share one property with cancer cells that makes them less than ideal as a model for studying the effects of oncogenes: like the ancient Greek gods, they are immortal.

Immortalized cell lines, like NIH3T3, are vital resources to biologists all over the world. They allow us to study cellular and genetic processes that would otherwise require the sacrifice of laboratory animals. The first immortalized cell line established in a culture was the HeLa cell. Virtually everyone who maintains cultured cell lines for their research has some experience with the HeLa line, which was derived from an aggressive cervical cancer isolated in 1951 from a patient named Henrietta Lack, a 31 year-old African-American mother of five. A sample of her tumor, which was removed for diagnostic purposes, was delivered to Dr. George Gey, the head of the Johns Hopkins School of Medicine tissue-culture lab at the time. Cells from the tumor began to divide almost immediately, and quickly adapted to life in an incubator, becoming the first continuous mammalian cell line ever established in long-term culture (there are thousands of different cell lines now). Gey named the cell line using the first two letters of Henrietta Lack's first and last names, and maintained her anonymity until he died in 1970. He even passed along a fake name, Helen Lane, to sidetrack people and protect the real patient's anonymity. The cells were an immediate scientific hit in 1951 and became the cornerstone for many research projects over several decades, including work on cancer and AIDS. The cell line was also used in the development of the Salk polio vaccine. Henrietta Lack died of her cancer only eight months after being diagnosed. Unbeknownst to her, the cancer cells taken from her body were being propagated in a culture, and their descendants continue to thrive in the laboratories of thousands of scientists around the world 50 years later. HeLa cells have even traveled to space, aboard the unmanned satellite Discover 17, to determine the effects of zero gravity on cell growth. HeLa cells grow so easily and prolifically in culture flasks that, if you are not careful during feeding time, they will easily contaminate other cell lines growing in your lab. However, despite the importance of HeLa cells to science, Henrietta Lack's unwitting gift to humanity has generated much controversy. In the mid-1970s, her name came up in a conversation between a relative of hers and someone who knew the true identity of the source of HeLa cells, an unethical breach of patient confidentiality. Her adult children were outraged when they learned that their mother's cells were still being kept alive without their consent, and that the family had not received some financial compensation.

Henrietta Lack has since been honored in film documentaries and government testimonials, and has earned the gratitude of three generations of scientists. The controversy has died down a bit, but there is no reason to expect her cells to expire any time soon. HeLa cells, and other cell lines derived from deadly cancers, should continue to survive indefinitely in a culture, centuries and millennia after the deaths of the bodies from which they were harvested, as long as there are people (or robots) around to feed them.

Because cellular immortality is an abnormal property displayed by cancer cells, biologists reasonably wondered about the relevance of the NIH3T3 transfection experiments. The question that needed to be addressed was, how would a normal cell respond to oncogenes?

The answer was established in an important experiment designed by Robert Weinberg's group at MIT, and separately by Earl Ruley, both reported in the same issue of *Nature* in 1983. Weinberg took normal cells-rat embryo fibroblasts-and sequentially introduced oncogenes that had been recovered by recombinant DNA techniques. At each step, they determined whether the engineered cells could grow into a cancer when introduced into athymic nude mice. A single oncogene did not suffice; but two did—and a third made the cancer more aggressive. This experiment, and many similar ones that followed, demonstrated that in order to convert a normal mammalian cell into a cancerous growth, at least two oncogenic abnormalities have to occur in the same cell; in other words, the conversion of a normal cell into a cancer is a multi-step genetic process.

This idea fit perfectly well with the clinical observation that exposure to a carcinogen does not immediately cause cancer, and that years may pass before the disease emerges. For example, workers exposed to asbestos in the Navy shipyards during World War II seemed unaffected until they developed very rare, asbestos-associated cancers 20 or 30 years later. Similarly, an increase in the number of leukemia cases occurred among residents of Hiroshima and Nagasaki after the atomic bomb attacks, but they only emerged years after the radiation exposure. The oncogene abnormalities induced by asbestos and radiation were necessary for cancers to develop, but were not sufficient on their own. Additional abnormalities had to accumulate over time in a damaged cell in order for cancerous transformation to occur.

The multi-step character of cancer-cell development keeps an even greater number of lives from being lost to cancer than already occurs. Every time a cell divides, it sustains hundreds and thousands of errors, an inevitable consequence of the replication mistakes made by DNA polymerase when DNA is copied (see chapter 1). Chromosomes are also vulnerable to breakage and recombination during replication, which can disrupt genes. Most of these changes are innocuous. However, with so many trillions of cells in the body undergoing cell division and acquiring spon-taneous mutations, it is inevitable that every now and then a mutation or chromosomal disruption will occur in a gene coding for a growth factor regulator that converts it into an oncogene. If a single oncogene abnormality was capable of transforming a normal cell, cancers would develop early in life, and often. Humans, with the very long stretch of time it takes to nurture newborns before they reach the age of reproduction, would never have survived as a species; everyone would die of cancer before reaching puberty. However, with at least two independent oncogene defects needed for cellular transformation to occur, not only do we have the capacity to survive to the age of reproduction, but there is also a fighting chance of our surviving into our 50s, 60s, or 70s before cancers begin to emerge as major health problems.

Nevertheless, our genome can only hold out for so long. About one in every two people who survive to the eighth decade of life will have had at least one bout with a cancer, and one-third will die of it, making cancer the second-leading cause of death in industrialized nations, after cardio-vascular disease. The dramatic increase in the human life span we have enjoyed in the last 75 years, through our ability to prevent and overcome deadly infectious diseases, comes with the price tag of an increased probability that several oncogene mutations will accumulate in a single cell that will turn it into a cancerous growth. Thus, spontaneous mutations and chromosomal alterations–the processes responsible for generating genetic diversity, which is the driving force of evolutionary change in bacteria and animals– also cause cancers that kill one of every three people in the industrialized world. It's the downside of DNA's error-prone chemistry.

Although there is an element of inevitability in the cancer equation, human behavior can affect the odds. The risk of acquiring an oncogene mutation increases with deliberate exposure to cancer-causing agents, such as cigarette smoke. Ninety percent of the lung cancers–the most common cause of cancer deaths–are caused by cigarettes. Applying sun screen can reduce exposure to DNA-damaging ultraviolet radiation and reduce the risk of developing skin cancer. Cancer risk can be reduced with proper diet: broccoli and other fresh vegetables contain natural anti-oxidant chemicals that reduce the effects of endogenous or exogenous, DNA-damaging oxidizing agents; and a diet rich in nuts, fruits, and vegetables and low in unsaturated fats has been shown to reduce the cancer rate in humans and test animals.

The dam was broken with the discovery of *src,* and a flood of oncogene discoveries has followed without interruption for more than two decades. In addition, the discovery of oncogenes also provided the intellectual framework that led to the identification of the other side of the cancer-causing equation. Unrestrained growth is not simply about cells gone wild because of oncogenic abnormalities that force the replication machinery into overdrive; scientists also found that cancers can be driven by mutations that disable genes, specifically growth-suppressor and apoptosis genes. Inhibiting a growth suppressor can have the same effect on a cell as activating a growth stimulator. Imagine a car getting wrecked (cancer) in a collision with a tree; the car will sustain damage if its engines are turned on without a driver (oncogenic abnormalities), or if the emergency brake malfunctions on a steep incline with the engine off (abnormalities in growth suppressor and apoptosis genes), or combinations of both. One of the most important tumor suppressor genes in the genome is p53. Among its many functions are a role in repairing mutated DNA and inducing damaged cells to undergo apoptosis; without p53, and similarly functioning genes, cells rapidly accumulate cancer-causing mutations. So important is a functional p53 gene in the DNA repair pathway that it has earned the lofty designation "guardian of the genome." Mutations that inactivate the p53 gene are now known to be the most common genetic abnormalities found in human cancer.

At the beginning of the twentieth century, the pioneering immunologist Paul Ehrlich used the term "magic bullet" to describe antibacterial substances that target specific microbes, analogous to the precise antigen-antibody interactions he had studied. A century later, the discovery of oncogenes has provided the opportunity to develop magic bullets for cancer that specifically target proteins encoded by oncogenes. The protein encoded by the *ras* oncogene, for example, is different from the one produced by the normal version of *ras*. This presents the possibility of finding a drug that will inhibit the mutant version but spares the normal one. Cancer therapy has been based on using drugs or radiation that target rapidly growing cells. These treatments are non-specific; cancer cells are killed, but so are normal cells, leading to the side effects commonly seen in patients receiving chemotherapy, such as loss of hair and damage to the gastrointestinal tract. "Designer" anti-cancer drugs that target defective proteins encoded by oncogenes should be more specific and effective, and cause less troublesome side effects. Although oncogene science is only two and a half decades old, the first set of drugs based on these discoveries is already in the treatment pipeline. One is imatinib mesylate (Gleevec is the trade name), made by the Swiss drug company, Novartis Pharmaceuticals. It's a drug designed to interfere with an abnormal enzyme produced in chronic myelogenous leukemia (CML) caused by the signature oncogene abnormality found in this bone marrow and blood malignancy, which is called *BCR-ABL*. The *ABL* portion of the gene codes for an enzyme called a tyrosine kinase, a key component in the signal transduction pathway that leads to DNA replication. Attachment of the *ABL* gene to *BCR*, coupled with the loss of *ABL's* first exon, leads to an increase in tyrosine kinase activity and an unwanted stimulation of DNA replication in a specific population of bone marrow cells, resulting in leukemia. The driving force behind the use of Gleevec in CML was Brian Druker a researcher at the Oregon Health and Sciences University. Alex Matter and Nicholas Lydon, researchers at Novartis, discovered the chemical inhibitor, but the company showed little interest in the drug because it caused liver damage in dogs. Druker was studying CML using both cell lines derived from patients with the condition and a mouse model of the disease-created by introducing the abnormal *BCR-ABL* oncogene into fertilized mouse eggs (the resulting embryos express the oncogene in bone marrow cells and develop a CML-type of leukemia). The drug proved to have a dramatic growth suppressant effect on his cultured cells and cured mice with CML, without any evidence of liver damage. But because of the toxicity problem in dogs, Novartis was still reluctant to release large quantities of the drug needed for clinical testing. However, Druker was so impressed with the laboratory results he took the unprecedented step of persuading fellow researchers to petition the company for the drug in order to conduct clinical testing. Novartis finally agreed to release Gleevec. The results of Druker's earliest clinical trials were striking, miraculous to some, especially when used in the early phase of the disease, where it induced a remission in 90 percent of test subjects without causing major side effects. The drug is a bit less effective in the more serious late phase of the illness. Although the long term beneficial effect of Gleevec on CML is not yet known–the drug has only been used for a few years–many patients seem to have been cured of the disease. Novartis was rewarded handsomely for supporting the basic and clinical research that went into the discovery of Gleevec, with earnings in the billions of dollars; the price tag for consumers (and their insurance companies) in need of this potential cancer cure-about $2,000 a month.

One predictable problem that has emerged with Gleevec, and one we can expect of other drugs like it, is the development of drug resistance caused by Darwinian selection, similar to the resistance that occurs when bacteria are attacked with antibiotics. There are so many hundreds of billions of cancer cells present in a patient with CML that spontaneous mutations are bound to occur in a few cells that will allow them to grow in the presence of Gleevec. Investigators studying Gleevec resistance in patients have identified, not too surprisingly, a number of mutations in the ABL portion of the BCR-ABL oncogene complex. These cause a change in the shape of ABL protein that interferes with the binding of Gleevec, allowing the cells to escape destruction. Thus, survival of the fittest is not simply a principle that affects entire living organisms, it can also explain some aspects of the growth of individual cells within an organism, with cancer cells behaving in some respects like autonomous living entities, like parasites, undergoing genetic changes that benefit their own survival at the expense of the host.

However, despite the problem of drug resistance, the successes achieved with Gleevec, and similarly designed treatments, for CML and other malignancies, show that anti-cancer magic bullets have indeed arrived. The lessons learned from the early experience with these drugs, as well as those gleaned from studying antimicrobial resistance and understanding survival of the fittest principles, suggest that the best way to effectively combat cancer cells will be for scientists to develop two or more specific treatments targeting different biochemical pathways affected by oncogenes, to reduce the chance that spontaneous mutations and Darwinian selection will provide the means for resilient cancer cells to escape pharmacological assault.

Thus, from an encounter between Peyton Rous and a lumpy chicken at the beginning the 20th century, to discoveries related to the isolation of reverse transcriptase in 1970 and the invention of recombinant DNA using parts stolen from the bacterial world, scientists have, with remarkable speed and efficiency, unraveled the cause of cancer in humans and developed the first specific anti-cancer treatments.

CHAPTER 16

Battle of the Genomes

"Man's evolutionary future biologically and culturally is unlimited."
—George W. Beadle

For families with genetic disorders, recombinant DNA has provided hope on two fronts: prenatal diagnosis and gene therapy. However, whereas for families who are known to be at risk, DNA-based prenatal diagnosis is now a clinical reality that enables them to have children free of lethal genetic disorders, the hopes for gene therapy have not quite lived up to its hype or people's expectations. Recently though, it seemed as if the 20 years of frustrating research into gene therapy might have finally paid off, after very dramatic successes in treating a severe inherited immune disorder, using retroviruses as vehicles for gene transfer were reported. But the ability of retroviruses to disrupt animal genomes turned the brief moment of success into a bit of a disaster.

Gene therapy was originally conceived as a way of correcting an inherited disorder by introducing a normal copy of the dysfunctional gene into the cells directly affected by the genetic problem. It has expanded to include novel ways to treat cancer and AIDS. In theory, gene therapy can be used to incorporate inhibitory genetic elements that incapacitate genes critical for cancer development or HIV replication. In fact, most of the gene-therapy clinical trials being conducted around the world are designed for treating cancer or AIDS. The potential market for these conditions is far greater than the market for rare genetic disorders, even though the rationale for–and feasibility of–using gene therapy to fight them are not as solid.

The primary method that researchers have explored in gene therapy for genetic disorders is to use recombinant DNA to engineer a corrective gene into a virus that can carry it into sick cells, where it will subsequently integrate into a chromosome and become a stable, expressed gene. This is a fairly simple idea to conceptualize, but in reality quite challenging (even Herculean, in the case of some diseases) to achieve. In treating sickle-cell anemia and thalassemia, for example, gene therapy would involve the introduction of a normal copy of the beta globin gene into most of the red blood cells in the body, through viral mediated transfer into nucleated bone marrow precursor stem cells, which is no mean feat. In treating cystic fibrosis, there are so many parts of the body damaged by the loss of CFTR (cystic fibrosis transmembrane regulator [see chapter 10]), it would be impossible to introduce a corrective gene everywhere. Researchers instead have focused on correcting the most damaging aspect of cystic fibrosis, the chronic bacterial infections caused by the loss of CFTR protein in airway epithelial cells.

A successful vector for gene transplant has to have a number of properties to make it an effective gene-carrying vehicle. First, for many genetic disorders, it must have the capacity to infect a large number of target cells. If only 1 percent of airway epithelial cells incorporated a normal CFTR gene through gene therapy, for example, it would do nothing to reverse airway disease. The virus should also be attenuated so that it cannot spread to other people or to other cells in the recipient's own body. The target cell should, if possible, include rejuvenating stem cells so that as mature cells age and die, the transplanted gene doesn't completely eliminate itself from the recipient. And finally, the transplanted gene has to be expressed at high enough levels to produce an amount of protein sufficient to correct the underlying physiological abnormality. These are all tremendous biological obstacles to overcome, and there is no single magic solution for all. But, for most gene-therapy researchers, retroviruses seemed to be the vector versatile enough to carry out most of these complicated demands. Retroviruses infect a large percentage of cells and integrate into chromosomes to form stable genetic structures.

Foreign genetic elements that remain in the cytoplasm are often destroyed by powerful enzymes called nucleases, so chromosomal integration is an important element in gene therapy, although it is also its Achilles' heel.

The first attempt to use a retrovirus for gene therapy in humans was a debacle, based more on the dream of a grand headline than on high-quality science. In 1980, Martin Cline, a talented hematology researcher from UCLA, attempted gene therapy on two children suffering from thalassemia. Since he failed to get permission for the venture in the United States (as the strategy for gene therapy was considered by the FDA to be too premature in its development to risk testing it on people), the experiment was conducted in Italy and Israel, where, at the time, human research protocols were more relaxed than they are today. Cline had argued, not without some rationale, that thalassemia was such a terrible condition that it was worth the risk of resorting to an untried therapy. The families agreed. But in 1980, the technology had not been developed sufficiently for such a grandiose scheme. In fact, the gene therapy trial was carried out in humans even before it had been perfected in laboratory animals, an unconscionable breach of standard scientific practice. Cline's talent and judgment may have been clouded by the prospect of becoming the first person to perform successful gene therapy. He was denounced by his peers and colleagues, was forced to resign as chairman of his department and lost his grants, and his research career was seriously damaged.

After nearly two decades of research, during which improvements in retroviral vectors were made and other methodological issues were perfected in rodents, gene therapy researchers finally scored a victory. A rare disease of the immune system called X-SCID was successfully treated in a few children. The life of children suffering from X-SCID is one of constant danger–a minefield punctuated by frequent bouts of serious, and ultimately fatal, bacterial and fungal infections. The adaptive immune system of such a child is completely ineffectual: neither the T-lymphoctyes nor the B-lymphocytes work, causing an immune deficiency worse than AIDS (which is "only" deficient in a subgroup of T-lymphocytes). The only way to prolong life in an X-SCID child is by a bone marrow transplant, if a compatible donor can be found. If not, that child has to be injected periodically throughout his life with gamma globulin, the antibody-rich serum fraction, a treatment that offers only partial protection against serious infections. These children are sometimes called "bubble boys" in the medi3a, because the abnormality is a mutant version of a gene called *IL2RG*, which is located on the X-chromosome, thus primarily affecting boys. *IL2RG* codes for a subunit of the receptor for the immune system activator, interleukin 2 (IL-2). Without it, lymphocytes cannot respond to the immune enhancing effects of IL-2. The "bubble" part of the term comes from the highly publicized life of a young Texan named David Vetter, who spent his twelve years of life housed in a germ-free plastic bubble.

In 1999, scientists in France reported the first unequivocally successful gene-therapy treatment. An engineered retrovirus containing a normal copy of *IL2RG* was introduced into bone-marrow cells of an affected child, and it corrected the immune deficit. Gene-therapy scientists and families with this and other genetic disorders were elated.

Then, two disasters struck. The first–unrelated to X-SCID– occurred in a gene-therapy trial in the year 2000 conducted at the University of Pennsylvania, where scientists were experimenting with an engineered version of adenovirus as a vector for carrying genes into cells. Adenovirus is a DNA virus that causes upper-respiratory infections in humans and has the capacity to infect a variety of mammalian cells. It grows as an autonomous DNA element outside the host cell nucleus and does not integrate into the genome, a characteristic that researchers are trying to exploit in creating a more ideal vector for gene therapy (for reasons that will become more apparent later in this chapter). The University of Pennsylvania scientists were testing different concentrations of adenovirus on volunteers to determine the maximum viral load that can be safely administered. One of the volunteers for the study was Jesse Gelsinger, an Arizona teenager born with OTC (ornithine transcarbamylase) deficiency, a relatively rare genetic disorder affecting the metabolism of urea and ammonia in the liver. Urea is a breakdown product of proteins. Untreated OTC deficiency causes seizures, brain swelling and coma, and can lead to severe mental retardation. OTC deficiency can sometimes be controlled with a low-protein, vegetarian diet, if diagnosed early in life, which was Jesse's situation. Since his problem was well controlled, Jesse would not necessarily have been a candidate for future gene therapy; but he volunteered for the adenovirus study anyway, because he wanted to participate in the development of a treatment that would help others who suffer from untreatable genetic disorders. He and several other OTC patients took part in a study in which the virus was infused directly into one of the main arteries leading to the liver. Although the other volunteers tolerated adenovirus infusion with little more than a temporary flu-like illness, Jesse's immune system revolted against the virus with an overwhelming inflammatory response that damaged his internal organs. Within a few days, he was dead, the first fatality in a gene-therapy trial. Deaths have occurred before in clinical trials without being reported to the public, but this one was highly publicized, given the circumstances and the pervasive fears of genetic manipulation that currently exist in society. A writer for the *New York Times*, Sheryl Gay Stolberg, referred to it as a "bio-tech death." Jesse's death temporarily shut down gene therapy research at the University of Pennsylvania. Investigators also found that leaders in other gene therapy trials around the country were under-reporting serious adverse effects from their human experiments. The aura of gene therapy, it seemed, had once again clouded the scientific judgment of some researchers. In response to these serious problems, the FDA placed a temporary ban on the use of adenoviruses in gene therapy. There was more bad news to come.

The researchers conducting the X-SCID gene therapy trial, which had been running smoothly for several years, with 10 boys successfully treated in France and four more in London, were caught by surprise when two young children in the trial developed leukemia within a few months of each other. A third developed leukemia several years later. It had been known since the discovery of oncogenes that retroviruses have two ways of causing cancer. One is to procure an oncogene, such as RSV (Rous Sarcoma Virus; see chapter 15), from its animal host. The other is by inserting itself into the genome near the normal cellular counterparts of oncogenes. Retroviral termini contain powerful gene-transcription promoters that can overactivate nearby genes, a phenomenon called insertional mutagenesis. Retroviral insertion can push nearby genes into overdrive, and if this is done to a gene that regulates growth, it can cause runaway cell division, a prelude to cancer. Retroviruses can also disrupt gene expression by inserting themselves into the coding region of genes. If the disrupted gene codes for a growth suppressor, excess cell division or loss of apoptosis can occur.

Insertional mutagenesis has been regarded, theoretically and experimentally, as being a random process. Disturbing the expression of genes involved in cell growth was expected to be a measurable but rare consequence of gene therapy using retroviral vectors, like a chance encounter with a neighbor while traveling in a distant foreign country. In fact, only a single episode of insertional mutagenesis was reported in thousands of subjects taking part in human and animal gene therapy trials using retroviruses. Yet, two out of the first 14 X-SCID gene therapy recipients came down with leukemia from retroviral insertional mutagenesis. Even more surprising was the finding that the engineered retrovirus had inserted itself next to the same gene, *LMO2,* in both children. *LMO2* is a well-known player in leukemia research circles; spontaneous disruptions of the gene were known to occur in T-lymphocyte leukemia found in the general population. Retroviral insertion in the X-SCID children was not as random as scientists thought. The FDA placed a moratorium on gene therapy trials using retroviruses inserted into blood stem cells until this latest setback can be examined thoroughly and some conclusions drawn. Thus, the promise of gene therapy seemed to come to fruition with X-SCID, but with a big asterisk.

The exalted highlight of the recombinant DNA era has been the Human Genome Project, the multibillion-dollar, decade-long effort to sequence the entire human genome, all three billion nucleotides of it–biology's big-budget equivalent to the lunar-landing program during the 1960s. The idea was first conceived in the late 1980s, when Robert Sinsheimer (from the University of California at Santa Cruz) and Charles DeLisi and David Smith (biologists at the Department of Energy) brought up the idea with prominent geneticists, and it soon developed a momentum all its own. One of the most vocal supporters was James Watson. But, there were prominent detractors as well. David Baltimore argued that blindly sequencing the genome violated the fundamental way that scientists tackle problems, which is from the bottom up, starting with individual, investigator-driven research. Sydney Brenner joked that the work would be so tedious and repetitive that prisoners should be assigned to the project as punishment– "the more heinous the crime, the bigger the chromosome they would have to decipher." However, by 1989, a plan was developed by genetics experts that satisfied most critics. The Human Genome Project would not simply be the blind deciphering of human DNA, but would transpire in well-planned phases of development that would yield important scientific and technical information all along the way, even before large-scale sequencing had commenced. There would be a focus on biological and computer technology, to improve the efficiency and reduce the cost of sequencing so many nucleotides, and to help decipher the enormous volume of data. These advances would benefit all members of the scientific community. There was also a plan to fund sequencing projects involving smaller genomes, as practice for the bigger, more ambitious human genome.

According to the initial timetable, the entire human genome would be sequenced within 15 years. A special office at the NIH was established for the initiative, the National Center for Human Genome Research, and in 1989 Watson was chosen to lead it. Watson's reputation, his enthusiasm for the project, and his ability to raise money made him an ideal choice to launch the bold scheme. The NIH and the Department of Energy would foot the bill on the U.S. side of the project. However, Watson quickly found himself at odds with NIH Director Bernadine Healy over their opposing views on the patenting of genes. Scientists who had cloned medically useful genes that were potential biological therapeutic agents, such as human insulin, had been granted patents for their discoveries. Patents were issued if the information derived from the characterization of an entire cDNA could be directly used to turn it into a useful therapeutic or diagnostic agent. On this point, there was some honest disagreement in the scientific community. However, Healy supported the idea of patenting genes that had only been partially sequenced, which provided only a limited amount of information that could not, on its own, be turned into anything of commercial value. These were being churned out at the NIH by Craig Venter, a former surfer and Vietnam medic turned biochemist, who had established a small DNA sequencing factory and was obtaining partial sequence information from thousands of genes, randomly picked from cDNA libraries. Venter infuriated the scientific community when he began, with the NIH's sanction and Healy's approval, to patent these partial sequences. It was a bit like an explorer/conqueror claiming a piece of foreign land based on spotting it from a distance at sea, without actually setting foot on its shores. Although Healy may have been more interested in testing the patent laws governing recombinant DNA-generated sequence data than in controlling the rights over genes discovered at the NIH, Watson (along with most of the scientific community) was vehemently opposed to the idea. Taken to an extreme, a researcher who fully characterized a gene that had commercial value could be sued by the "owner" of the partial gene sequence. Watson has the refreshing but risky habit of speaking his mind–in private conversation, in public lectures, and his writing–and he freely expressed his disgust with Healy's view on patents. Healy countered that Watson was a poor choice to lead the Human Genome Initiative because his ownership of biotechnology stocks created a conflict of interest. In April 1992, Watson resigned before "that woman" had a chance to fire him. Michael Gottesman was appointed interim acting director.

The search for a new director did not last too long. In April 1993, Francis S. Collins was appointed to take over the project. He turned out to be the perfect choice. Watson's vision and energy were important to get the sequencing project off the ground, but Collins's calmer, less cantankerous voice was needed to ensure that the project ran smoothly. There were too many personalities and egos to sooth, and too many critics to assuage. Collins, a born-again Christian who had been home-schooled by his mother through the sixth grade, was a leading genetics researcher and one of the principal investigators involved in the successful hunt for the cystic fibrosis gene.

It was clear from the beginning that sequencing the human genome would require automation. Standard sequencing methods were too slow and expensive. Most researchers used a method developed by Fred Sanger, called dideoxy termination, a very clever variation of the standard DNA replication reaction, the type that takes place whenever dividing cells double their DNA. It turned out to be perfectly adaptable to automation. During the process of replication, two complementary, single-stranded pieces of DNA incorporate nucleotides according to the A:T, G:C rule to create two exact copies of double-stranded DNA (every A nucleotide on one strand of the double helix binds to a T nucleotide on the other strand; and every G binds to a C). However, for a second DNA strand to grow based on the nucleotide sequence of the first, an inserted nucleotide not only has to form the appropriate base-pair connection to its partner on the opposite strand, it also has to form a chemical interaction (called a phosphodiester bond) with the nucleotide in front of it and to the next one coming in. Visualize two long chains of people facing each other with the front of their bodies touching (A:T, G:C base pairing), each holding hands with the person on either side of them (phosphodiester links). Sanger figured out a way of interrupting this process in an orderly and controlled manner, using nucleotide derivatives called dideoxynucleotides (chain terminators), which are able to form a phosphodiester linkage on one side of the chain but not on the other. These nucleotides can form A:T, G:C pairings and bind to a growing DNA chain on one side, but cannot accept the next nucleotide, thus terminating DNA synthesis at that site. Picture the long chain of hand holders being interrupted by a one-armed person.

By using experimentally determined ratios of regular nucleotides and chain-terminating dideoxynucleotides, Sanger was able to generate a series of DNA molecules in a test-tube replication reaction in which a dideoxy terminator was randomly inserted into a growing DNA molecule. Using four separate reactions, each containing either an A, T, G, or C terminator, Sanger was able to generate about three hundred DNA molecules that differed in size by a single nucleotide, each stopping at an A, T, G, or C, depending on which dideoxynucleotide was used in a chain-termination reaction. These could be separated from each other by gel electrophoresis, which separates molecules in an electric field according to size. If a small amount of radioactivity was included in the reaction, each molecule could be seen on an X-ray film. By loading each chain terminator reaction separately onto the gel, the hundreds of molecules would appear as an irregular stepladder, one rung corresponding to a chain-terminated nucleotide at one position, its next-door nucleotide rung in another. These corresponded to the order of nucleotides in the DNA molecule being replicated. In other words, if nucleotide number 100 in a chain of DNA ended with a T terminator, there must be a "T" at position 100 of the strand being replicated and an "A" on the corresponding complementary strand. If nucleotide 101 ended with a "G" terminator, there must be a "G" at the site and a "C" on the complementary strand, and so on. Using this strategy, approximately three hundred nucleotides of DNA could be deciphered in a single reaction. The process was fairly tedious; each reaction had to be done four times (one for each nucleotide), gels for electrophoresis had to be prepared, and data was processed by eye. To reduce errors, both strands of DNA had to be sequenced. Then, in order to sequence an entire cDNA, which typically contains several thousand nucleotides, you had to do it all over again, 10 or 15 times for each strand, until every nucleotide was accounted for. All in all, it was slow going.

Sanger's lab used his technique to sequence the first complete organism, a small bacteriophage called phiX174, containing about 4,500 nucleotides–a process that took a couple of months to complete. Although Sanger's Nobel Prize-winning dideoxy invention was simple–an extremely clever variation of a normal DNA replication reaction–the strategy, as originally conceived and then used by researchers all over the world, would never work well for the human genome project. It was just too slow. Sequencing the genome by hand would be like building the pyramids with only one laborer. With three billion nucleotides to sequence, it would take one person about 20,000 years, working round the clock, to complete the project; a thousand people could do it in 20 years of full-time work, if they took no holidays, vacations, or sick days, and gave up sleep. The cost in labor and reagents would be outlandish, and the process would indeed be punishingly tedious. It was clear that in order to read the human genome and manage costs, a faster and more efficient method of sequencing was needed.

The two major innovations applied to the Sanger method that made whole genome sequencing possible were the use of flourescent dyes as a substitute for radioactivity to label DNA, and automated DNA-sequencing machines to read them. Building a machine to read DNA sequences was the brainchild of immunologist Leroy Hood and his Cal Tech colleague Michael Hunkapiller. By using flourescent dyes that emitted different colors for each dideoxy nucleotide reaction, the replication reactions that generated the sequence ladder could be accomplished in one tube instead of four. A laser beam would excite the flourescent dyes to emit a small amount of color, which, after being focused by a lens onto a photomultiplier, could be digitalized and analyzed by computer. The DNA sequence reactions were separated on pre-made gels cast in thin tubes, which increased resolution and the number of bases that could be analyzed in one reaction. Soon after coming up with the concept of the process, Hood teamed up with an instrumentation company to found Applied Biosystems Incorporated (A.B.I.), and within a year, a working prototype was built. Instead of a few hundred nucleotides a day, scientists could now ramp up DNA sequencing to thousands a day, all in a color-coded format that could be transferred as a computer file for very easy handling and analysis. Using several sequencing machines, millions of nucleotides could be sequenced in a matter of weeks and months. With hundreds of sequencers, anything was possible, even sequencing the three-billion-nucleotide human genome. The only drawback for the average researcher was the cost-sequencing machines were priced at roughly a quarter of a million dollars in those early days. Most research facilities ended up purchasing one or two for the entire institution, and they had to be shared by everyone who sequenced DNA.

Another issue that had to be tackled by Collins and other scientists researching the human genome was whether to sequence the entire genome, junk and all, or simply focus on the coding regions and their immediate environs, which would provide the greatest scientific return for the dollar. There were advocates for the latter approach, but Watson and Collins both favored doing it all, which greatly magnified the scope and cost of the project.

They reasoned–correctly, it turned out–that the only way to understand human evolution was to do a complete comparison with its mammalian cousins. They also reasoned, again correctly, that hidden functions would be found in the junk that would become apparent after the project was completed.

Finally, there was the question of how to assemble all the pieces together. In order to sequence a large fragment of DNA, it has to be broken down in smaller pieces and cloned in bacteria or yeast, with each piece sequenced and carefully put together, a slow and methodical (but proven) method favored by most geneticists. The other approach, developed by Craig Venter and Hamilton Smith, the discoverer of restriction endonucleases, was the "shotgun" method, where DNA would be sequenced randomly from DNA libraries, their overlapping ends matched later using powerful computer algorithms, like piecing together a shattered piece of pottery using a computer capable of recognizing the ends of individual shards, figuring out exactly where each piece belongs. The shotgun idea was not well regarded by most geneticists, who argued that genomes were too complex to obtain an orderly reading of the sequence from A to Z with this method. Too many of the broken shards looked alike for the broken pottery to be properly reassembled, it was argued. But Venter was convinced he could do it.

He had spent a year at the NIH, tediously using the conventional sequencing approach just isolating and decoding a single gene. When he read about Hood's work with automated sequencing, he quickly realized that this was the solution to his problem. Venter was able to convince the NIH to procure an A.B.I. sequencing machine, becoming one of the first researchers to use the device. Soon, he was sequencing randomly like mad from cDNA libraries, creating the thousands of gene fragments that provoked the patenting controversy between Watson and Healy. However, Venter had a bigger vision than blindly sequencing partial gene fragments from cDNA libraries and arguing about patent rights. It was whole genomes he was after, and he would start first by cracking the code of a bacterium, which had never been done. With thousands of genes, the bacterial genome would dwarf the dozens of genes present in the viruses that at that point been the only organisms completely deciphered. But Venter realized that even with automation, a bacterial genome would take years to unravel. He and Smith came up with the shotgun approach as a way of speeding things up and saving money. However, an application to the NIH to pursue the idea was rejected by proposal reviewers there; the shotgun method was deemed too radical.

To follow through with his dream of sequencing an entire bacterial genome, Venter had to leave the NIH and start his own company. With money from venture capitalists, who in the early days of the bio-tech frenzy were throwing money around to fund almost any project that scientist/entrepreneurs pitched, Venter started a gene-sequencing company, The Institute for Genomic Research (TIGR). He convinced the money men that the company would earn income, by patenting medically and scientifically useful products, and through peddling information by charging a fee to scientists and companies interested in early access to sequencing data. Drug companies with access to bacterial sequence information might use the data to create novel antibiotics, for example. With dozens of DNA-sequencing machines available, Venter created a gene-sequencing factory on a scope never before assembled under one roof, and stunned the genetics world in 1995 when he finished the entire sequence of that first bacterium in a mere nine months.

The bacterium was *Hemophilis influenzae.* Consisting of a total of 1.8 million nucleotides, it was the largest completed sequencing endeavor in history, dwarfing the bacteriophage- and viral-sequencing projects that had been accomplished up to that time by almost two orders of magnitude. Hamilton Smith had recommended *Hemophilis influenzae* for the first bacterial genome project, the organism used for his Nobel Prize-winning isolation of the first restriction endonuclease. It was the first completed sequence of a free-living organism (viruses are intracellular parasites, and cannot survive on their own). In choosing *Hemophilis influenzae,* they bypassed the scientifically more important *E. coli*, which other groups had been struggling to sequence for nearly a decade. In the time it took for a graduate student or postdoc to clone and sequence a single bacterial gene, Venter's factory had sequenced the whole lot of them, 1,700 in all. It was a spectacular victory for the shotgun approach, and a slap in the face to his detractors. Soon, other bacterial genomes were deciphered by TIGR, and the race began to determine which non-TIGR facility would be able to complete a bacterial sequencing project of their own.

Geneticists had to admit that the shotgun approach had worked beautifully in sequencing bacterial genomes, but most were still resistant to the idea that complex mammalian genomes could be reconstructed in the same manner. The abundance of repetitive DNA created too many pieces that looked precisely the same for any computer program to reassemble in an orderly fashion without error.

By the mid-1990s, Collins had assembled a group of top geneticists from the United States, England, Japan, France, and China, about 20 groups in all, to work on the Human Genome Project. Each was assigned the task of sequencing an entire human chromosome, which range in size from about 20 to 200 million nucleotides. The U.S. contingent was led by Eric Lander, a Rhodes Scholar mathematician turned geneticist, and director of the Whitehead Institute. The money and facilities were in place; the sequencing technology had been perfected. Collins even had the insight and political savvy to devote a small percentage of the billion-dollar budget to the social and ethical implications of the genome project to handle important issues such as genetic privacy and genetic discrimination in health insurance and hiring. The eventual upshot of this aspect of the genome project–to which many scientists were opposed, viewing it as a waste of resources–was the enactment of legislation adopted by most states designed to protect private genetic information. It also appeased civil libertarians, bioethicists, and other critics, who were also concerned, rightly so, about the impact the human genome project might have on the possibility of creating genetically engineered humans, or selecting "perfect" embryos. Underlying some of the concern though was the fear that the human genome sequence would provide proof of an undemocratic genetic world, the unequal distribution of the assets and flaws individual genomes had to offer.

The project was moving along, slowly, but moving along. The official target date for completing the project was 2005. However, Collins and the rest of the consortium were forced to reassess their timetable when Venter announced that he would personally beat their target date and have a completed sequence of the human genome by 2002. Venter had started a new company in 1998, Celera Genomics, for the express purpose of sequencing the entire human genome. Venter, with new venture-capitalist money amounting to nearly a third of a billion dollars, was able to purchase a small fleet of A.B.I. sequencers, 300 in all, and hire top geneticists and computer specialists. He announced that he would beat the consortium with his shotgun approach, using one of the world's largest supercomputers to put the millions of pieces of sequence data together, and make money to boot, by marketing human DNA sequence information to drug companies and other scientists. The pharmaceutical industry was interested in the genome as a huge potential source of information about genetic disorders to target for new drug development. They were particularly interested in G-protein coupled receptors, neurotransmitter transporters, and enzymes, all prime targets for a host of medications already on the market, from antacids and antidepressants to Viagra. In addition, the new field of pharmacogenetics was emerging: a future of personalized medicine based on genetic profiles that would provide useful clinical information about patients, such as predicting whether or not a therapeutic response or a nasty side effect will occur with a particular class of medications. A profile that also includes individual assessment for the risk of developing particular diseases, such as Alzheimer's and osteoarthritis, would also be part of an all-encompassing personal genetic package. A number of bio-tech and pharmaceutical companies were interested in developing and selling genome-related products, and were willing to buy sequence information to get a head start on the competition. The human genome was viewed as a commodity– Celera's very raison d'être.[40]

Collins and the project's group leaders were incensed. Not only did they see themselves getting beaten by Venter, which would be a bitter pill indeed, but they were troubled by the idea

---

[40]Gene-chip technology—the ability to load tens of thousands of DNA molecules on a slide, originally developed by the company Affymetrix—is making individual genetic profiling of health information a reality.

that, rather than serve the public good, the sequencing of human DNA could be held for ransom by a private company for its own profit. For years, the members of the consortium had argued and disagreed about numerous issues, often bickering about which course of action to take. But they all agreed on one thing–that the information obtained from the Human Genome Project (which was being funded by taxpayer money through the NIH and by private medical research charities such as the Wellcome Trust, a British philanthropy) should be made public, freely available to all scientists for their research. The Human Genome Project's record of posting huge blocks of sequence data for public consumption as rapidly as possible, basically from the sequence readout on a computer screen to an open Internet site, was unmatched in the annals of competitive science.

With Celera completing sequences by leaps and bounds, there was nothing Collins could do except step up his groups' rate of production another notch. In 1999, only 10 percent of the genome had been sequenced. The preceding years had been devoted to perfecting technology and DNA-reading computer software, getting all the cogs in place for the final sequencing campaign. The members of the consortium were instructed by Collins to complete their chromosome sequences, and fast. Sequencing commenced at an unimaginable clip, at one point averaging, collectively, nearly a thousand DNA bases a second, around the clock, from all over the world, a spectacular 80 to 90 million bases a day. Within months, sequencing of the two smallest non-sex chromosomes (numbers 21 and 22) was completed, and within a year, there was a working draft of the entire genome. It was not a complete, finished project. Because of the large amount of repetitive DNA in the genome, some ends had not matched up and some could not be cloned and sequenced at all; repetitive DNA is very difficult to grow in microbes. These final steps would require more care and time. But a 90 percent genome was a great start, and could be used almost immediately by investigators interested in the genetic basis of some complex medical disorders, such as diabetes and schizophrenia. Meanwhile, Celera was on track as well. But the competitive strain was showing on both sides. Collins and Venter decided to call a temporary truce and claim mutual victory. In a very public announcement in June 2000, attended by President Clinton and British Prime Minister Tony Blair, they declared the joint completion of working drafts of the human genome. The findings were published in February 2001 in the journals *Nature* and *Science*.

However, the truce between Celera and the consortium didn't last long. Collins and other genome scientists accused Venter of using freely available sequence information that had been posted in the public domain to help plug up some gaps in the Celera shotgun sequencing data. Venter claimed that his data were more accurate, since his speedier shotgun method had allowed him to sequence the entire genome several times over, reducing error.

In 2002, Venter left Celera. The company had decided to make drug discovery and development their main objective, which Venter regarded as a practical but mundane undertaking, causing him to no longer fit in with the company's business plan. The Human Genome Project, by posting DNA sequence data for free, had made it unnecessary to purchase human genetic sequencing information from Celera.

After resigning from Celera, Venter became the director of the Institute for Biological Energy Alternatives, an enterprise funded by the Department of Energy, charged with the mission of creating synthetic microbes that can be used to clean up waste products created by fossil fuels and convert them to usable energy. The idea is plausible, but it has a bit of twisted logic to it: create a possibly risky technology to clean up the mess of another. As a practice venture, he and his colleagues synthesized phiX174, the virus that was the first organism ever sequenced, thirty years ago; the little virus had been completely decoded at one end of the biotech spectrum, and completely reconstructed at the other end. Venter's synthesis of it from "off-the-shelf" nucleotides took only a few weeks, faster than nature's ability to convert chemical nucleotides into a life code by perhaps a billion years, although Venter's controlled laboratory conditions were more ideal than the dilute, sloppy stew of ingredients that were present in the primordial soup. Ever-restless, Venter has recently turned to randomly sequencing microorganisms in seawater. His analysis showed that the diversity of microbial life in the Earth's waters is greater than previously thought. Microbes that escaped detection previously because they were difficult to culture could not escape Venter's DNA sequencing machines. He is now trying to determine the diversity of microbial life by sequencing particulate matter taken from New York City air.

Meanwhile, the consortium quietly and successfully, without much public fanfare, put the finishing touches on the human genetic plan.

The human genome was found to be simultaneously large and small. Although it was big enough to require the equivalent of thousands of average-size books to house every nucleotide character in the genome, it ended up having fewer genes than imagined—essentially the same number as a rodent's genome, and barely twice as many as a fruit fly's. Before the genome project was completed, an informal running pool was set up by scientists to guess the total number of genes that would ultimately be found. Most guessed between 50,000 and a 100,000, but fewer than 30,000 genes have been detected. Genes are identified by computer programs that sift through the mountain of sequence information, recognizing genes by various signatures. The millions of different G-protein-coupled receptors that exist in the animal kingdom, for example, can each be recognized by their unique structural motif, an assembly of fat-soluble amino acids that enable them to traverse cell membranes seven times. Most other genes are members of families that can be recognized by their shared structural domains. It is like determining that an animal is a member of the cat family by viewing a picture of just a paw. The gene-identification programs are not perfect, so the computers that sift through the human genome miss a few genes, and there may be another few thousand more waiting to be revealed by improved computer techniques or even by the old standby method: individual, hands-on research. Yet, the relative paucity of genes in the human blueprint underestimates the complexity of our genome. Many genes in the genome, nearly 50 percent, have alternative splicing mechanisms in place that create more than one protein from the same set of coding elements. Most alternatively spliced genes code for several different (albeit related) proteins. In fact, some genes have so many exons that fully exploiting their alternative splicing potential would create thousands of different proteins. There is also an entire family of genes that code for RNAs that are not translated into protein, which are called non-coding RNAs. Non-coding RNAs influence the expression of other genes. Thousands of them probably exist in the genome, but were missed in the initial computer searches.

Geneticists are also realizing that the size of the genome is only one element in the complexity of life's construction plan. Physical structures and behavior can be altered by the interaction between genes and environment, and by the timing and magnitude of gene expression. For example, activation of growth-hormone expression in the first 15 years of life turns children's bodies into adult-sized ones, while the same hormone expressed after maturity grotesquely alters the face, hands and feet. Exposing rat pups to the stress of isolation from their mothers for a few hours after birth causes permanent changes in the brain's stress-response genes that impairs their exploratory behavior as adults. And inappropriate expression of genes involved in cell growth can transform a cell into a cancerous growth. When it comes to gene expression, timing is critical. There are also layers of complexity created by protein-protein interactions. Almost every protein has some effect on the function of other proteins. A transcription factor, for example, capable of stimulating the expression of dozens of different genes, can be completely blocked by binding to another transcription factor. Thus, the addition of a single gene to an organism's genome does not simply increase the weight or value of that genome by a single unit, its effect is amplified many-fold by the interaction its encoded protein has on other proteins. Even changing a genome by a simple mutation in a promoter region, which does not alter the number of genes in that genome, could have a drastic effect on the organism. A mutation that increased the expression of a gene critical for brain development, for example, could have as great an effect on brain complexity as adding on another gene. The differences between humans and chimpanzees can be explained more readily by such subtle genetic variations than by counting the actual number of genes in our respective genomes.

The human DNA sequencing data can be viewed at the Web site of the National Center for Biotechnology Information, although one has to know how to navigate the nucleotide alphabet to obtain useful information from it. You would have to scroll through the sequence for a year at a rate of one page of information every ten seconds to view the entire array of nucleotides. Other than the sheer magnitude and wonder of the data, the endless array of A's, T's, G's, and C's would mean, to most people, little more than the O's and 1's in a computer program.

In October 2002, reports of the complete DNA sequence of *Plasmodium falciparum* and its mosquito carrier, *Anopheles gambiae,* were published in the journals *Nature* and *Science.* More than two hundred scientists are listed as co-investigators, including Venter, who participated on both projects. In the issue of *Science* that reported the sequencing data for the mosquito, an evolutionary comparison was made with the fruit fly *Drosophila melanogaster,* whose sequence had been determined a couple of years earlier. Comparative genetic analysis showed that the two species evolved from a common ancestral fly about 250 million years ago. Quite an advance in understanding the evolution of a species. Yet, published in the same issue was a short news item reporting the Cobb County Georgia decision to allow the use of science textbooks that provide equal time to the alternative views of evolution as imagined by proponents of Creationism, a faith-based interpretation of the biological world. The amendment was adopted in a unanimous 7–0 vote by the local school board.

These nucleotide scripts that have been sequenced provide the code for thousands of different genes in three players that have participated in an endless drama over eons–humans, a species of mosquito, and a family of parasitic protozoa. Buried in the codes is the information defining all aspects of malaria transmission and disease, and the human response: the unique shifting chemistry of *Plasmodium falciparum's* cell membrane proteins that allows it to evade the human immune system; its ability to invade red blood cells, which provides it with nourishment and safe haven against antibody attack; the genes that are activated to propel the parasite into its critical life-cycle phases; the mosquito's attraction to human blood and sweat; its maternal drive to seek nutritional support for egg laying, risking its life by seeking out Earth's dominant species as a food source; the nesting and nocturnal behaviors that govern mosquitoes' biting patterns; the biological clocks that time the reproductive cycle of *Plasmodium* with the feeding behavior of mosquitoes; the genes for T- and B-lymphocyte diversity that ultimately fight the elusive and deceptive parasite; the protective mutations that have provided an extra shield for some people. All of these are factors involved in the point-counterpoint evolution of a disease that has killed billions of people, most before reaching the age of puberty.

The *Anopheles gambiae* genome has nearly 250 million nucleotides and 14,000 genes. *Plasmodium falciparum* has 23 million nucleotides and about 5,300 genes. As I write this, laid out in front of me are the issues of the journals *Science* and *Nature* showing the chromosomal positions of every gene in *Homo sapiens, Anopheles gambiae,* and *Plasmodium falciparum* that were identified in the different sequencing projects. Is there a magic bullet hidden somewhere on these gene maps that will finally put an end to the seemingly endless battle against malaria? One possible outcome of the *Anopheles* sequencing project is to help understand the relationship between mosquito and *Plasmodium* in order to develop strains that are resistant to parasitic infestation, so that for humans who still got bitten, there would be no *Plasmodia* in the insect's spit. One group of researchers has developed an *Anopheles* strain engineered to produce an antiplasmodial substance so that it cannot function as a vector for malaria transmission; whether it can compete with native *Anopheles* in the real world remains to be seen. This approach would be better than altogether eliminating disease-bearing mosquitoes by insecticides, which, even if it could be accomplished without harming the environment, might have a damaging effect on the lower-animal food chain. Another tack will be to identify genes involved in mosquitoes' biting behavior in order to help build traps. Mosquitoes are attracted to heat, water vapor, carbon dioxide, and volatile oils emitted from human skin. If all the attractant molecules could be harnessed together, a trap could be developed that might draw hungry mosquitoes away from humans. Even before the *Anopheles* sequencing project was completed, one company began marketing the "Mosquito Magnet" (costing $500 to $1,000), a trap that contains 1-octen-3-ol, a volatile attractant, along with water vapor and carbon dioxide, all heated to comfortable human body temperature. The *Anopheles* genome project will certainly identify other attractants. Analysis has already revealed 79 odorant receptors and 76 gustatory (taste) receptors, members of the family of G-protein coupled receptors. The most interesting ones to malaria researchers are those that are expressed only in mature females, making them good candidates for regulating feeding behavior, and targets for drugs that might diminish the female mosquito's unique attraction to humans and human blood. Researchers at Yale and Vanderbilt recently identified a receptor in *Anopheles* that binds to a chemical called 4-methylphenol that is secreted in sweat. It was identified by engineering the receptor gene into neurons of fruit flies. Chemicals found in sweat were added to the engineered neurons, but only 4-methylphenol activated the receptor. When looking for a sensitive detection system, scientists often do not have to look much farther than the neurons of their favorite creature for lab research, which have been designed by evolution to respond to minuscule concentrations of stimuli with an electrochemical signal that can be amplified and measured by modern electronics.

Analysis of the *Plasmodium* sequence may identify enzymes that could be attacked pharmacologically. Enzymes are particularly good targets for drug development because some are unique to the parasite and have no counterpart in humans, which theoretically should limit side effects. In 1999, before the publication of the *Plasmodium* genome sequence, a German group found an enzyme involved in *Plasmodial* fatty-acid synthesis that is inhibited by the drug fosmidomycin, chemically creating a lethal metabolic deficiency; clinical trials of the drug are near completion. More studies of this type are bound to increase in the near future, using analysis of the sequence data. The *Plasmodium* genome project also unearthed a host of surface-protein genes that could function as target antigens for a recombinant DNA vaccine.

The *Plasmodium* and *Anopheles* genome projects were not universally applauded by malaria researchers, some of whom claimed that the money granted for those efforts was diverted from other productive research projects. It was also argued that many potential vaccine antigens were already available from other research studies that had not yet been tested because of a lack of funds. They have a point. As it stands today, the total amount of money designated for malaria research around the world in 2002 was only $200 million. So, the efforts of the mosquito and parasite genome projects will amount to little unless follow-up projects, especially for a malaria vaccine, are adequately funded.

There is also a cautionary note in the hoopla over genomes. Partial resolution of the problem posed by malaria–by means such as the release of engineered *Plasmodium*-resistant mosquitoes into heavily malarious areas, which could reduce but not necessarily eliminate disease-carrying females–may lead to a less-than-ideal solution. As the number of malarious mosquitoes decrease, there is less local immunity (see chapter 3), which creates an environment favorable to a deadly upswing, should a small population of malaria-transmitters persist. The solution to silence "the whine of *Anopheles gambiae*" and its "music of death,"[41] will probably only come with a protective vaccine.

The human genetic code, with the instructions to synthesize every protein in the body, is stashed away in computers at the National Center for Biotechnology Information, the digital code of O's and 1's carrying the genetic code–millions of Rosetta stones waiting to be deciphered. However, human shortcomings are preventing us from taking full advantage of our technological virtuosity. The gains we have realized on one end of the human performance spectrum are matched by our underachievement on the other end. What good is a malaria vaccine if genocidal civil war in an affected country prevents safe access of health-care workers to cities and villages? How can polio be entirely eliminated from the world if pockets of disease persist in regions plagued by rumors that the vaccine causes sterility? George W. Beadle, in the same essay from which the quote introducing this chapter was taken, wrote: "But knowledge alone is not sufficient. To carry the human species on to a future of biological and cultural freedom, knowledge must be accompanied by a collective wisdom and courage of an order not yet determined by any

---

[41]Tom Clarke, *Nature*, no. 419 (2002):429–30.

society of man." The flaws in our psychology–propensity for violence, intolerance of different cultures, greed, graft, and selfishness, blind acceptance of tyrannical demagogues–that are responsible for so much misery in the world are creating conditions and draining resources that prevent us from tackling diseases to our fullest potential, and controlling the infectious microbes that are the enemies of us all. These traits made the twentieth century the most violent in human history, with the slaughter of more than 150 million people—and the twenty-first century hasn't began any differently. Studies in behavior genetics show that nearly every human personality trait is strongly influenced by genes; for many, genetic factors are as important as the environmental ones in the nature/nurture mix that controls the behavioral engine. The gene/environment interactions that help shape personality are enormously complex, but it's safe to assume that genetic factors play a key role in the proclivity humans have for destroying each other. While historians and political scientists properly explore the economic, political, and social aspects of violence and exploitation, the genetic factors that play a role in these behaviors and the evolutionary drive that maintains their existence in the human genome have largely been ignored.

So, there is much more to explore in the human genome beyond searching for new drugs, identifying genes that increase disease susceptibility, and establishing evolutionary comparisons with other species. The genome data is also offering us an opportunity to determine the genetic factors that play a role in brain development and behavior. Somewhere in the mix of *Homo sapien*'s three billion A's, T's, G's, and C's are combinations that contribute to good or evil. In the final analysis, we will probably find that the most unyielding and dangerous battle of the genomes, the one posing the greatest threat to human existence, is not being waged with infectious microbes, but the one we are waging against ourselves.

References Cited

Introduction

Brandtzaeg P (2003) Mucosal immunity: integration between mother and the breast fed infant. Vaccine 21:3382-3388.

Guyer B, Freedman MA, Strobino DM, Sondik EJ (2000) Annual summary of vital statistics: trends in the health of Americans during 20th century. Pediatrics 106(6):1307-17.

Chapter 1: The Microbial World

Beaber JW, Hochhut B, Waldor MK (2004) SOS response promotes horizontal dissemination of antibiotic resistance genes. Nature 427:72-74.

Coyne MJ, Reinap B, Lee MM, Comstack LE (2005) Human symbionts use a host-like pathway for surface fucosylation. Science 307:1778-1781.

Llosa M, et al (2002) Bacterial conjugation: a two-step mechanism for DNA transport. Molecular Microbiology 45(1):1-8.

Loeb M (2004) A shot in the arm. Nature 431;892-893.

Osborn AM, Böltner D (2002) When phage, plasmids, and transposons collide: genomic islands, and conjugative and mobilizable-transposons as a mosaic continuum. Plasmid 48:202-212.

Weigel LM, et al (2003) Genetic analysis of a high-level vancomycin resistant isolate of Staphylococcus aureus. Science 302:1569-1571.

Wenzel RP (2004) The antibiotic pipeline-challenges, costs, and values. New England Journal of Medicine 351(6)523-525.

West NP et al (2005) Optimization of virulence functions through glucosylation of shigella LPS. Science 307:1313-1317.

Chapter 2: Magic Bullets

O'Neill LAJ (2005) Immunity's early-warning detection system. Scientific American. January 2005, 38-45.

Hendesson B, Wilson M, McNab R, Lax AJ (1999) Cellular Microbiology: Bacteria-host interactions in health and disease. John Wiley and Sons, Chicister.

Marsh Steven GE, Parham Peter, Barber Linda D (2000). The HLA facts book. Academic Press.

Schwartz RS (2004) Paul Ehrlich's magic bullets. New England Journal of Medicine 350 (11):1079-1080.

Tauber AI (2003) Metchnikoff and the phagocytosis theory. Nature Reviews Molecular and Cell Biology. 4:897-901.

Chapter 3: The Mosquito Plague

Campbell P, Butler D (2004) Malaria. Nature 430:925-945.

Desowitz Robert S (1991). The Malaria Capers, W.W. Norton and Co. New York, London.

Hawking F (1970) The clock of the malaria parasite. Scientific American 222(6):123-131.

Ito J, Ghosh A, Moreira LA, Wimmer EA, Jacobs-Lorena M (2002) Transgenic anopheline mosquitoes impaired in transmission of a malaria parasite. Nature 417:452-455.

John Cornwell. Hitler's Scientists. Viking Press, 2003.

Ranson H, Claudianos C, Ortelli F, Abgrall C, Hemingway J, Sharakhova MV, Unger MF, Collins FH, Feyereisen R (2002) Evolution of supergene families associated with insecticide resistance. Science 298:179-181.

Renaud Lacroix, Wolfgang R. Mukabana, Louis Clement Gouagna, Jacob C. Koella (2005) Malaria Infection Increases Attractiveness of Humans to Mosquitoes. PloS Biology 3(9): e298.

Rocco (F) (2003) The Miraculous Fever-Tree: Malaria and the Quest for a Cure That Changed the World Harper Collins, Great Britain.

Warrell David and Giles HM, ed. (2002) Essential Malariology. Arnold Publishing Company London, New York, New Delhi.

Chapter 4: Killer Beans

Beutler E (1996) G6PD: population genetics and clinical manifestations. Blood Rev 10(1):45-52.

Kwiatkowski DP (2005) How malaria has affected the human genome and what human genetics can teach us about malaria. Am J Hum Genet77(2):171-92.

Meletis J, Konstantopoulos K (2004) Favism-from the "avoid fava beans of Pythagoras to the present. Haema7(1):17-21.

Ruwende C, Hill A. (1998) Glucose-6-phosphate dehydrogenase deficiency and malaria. J Mol Med76(8):581-8.

Vennerstrom JL, Eaton JW (1988) Oxidants, oxidant drugs, and malaria. J Med Chem 31(7):1269-77.

Chapter 5: Neither Hurry, Curry nor Worry: Blood Type and Disease

Aspholm-Hurtig M et al (2004) Functional adaptation of BabA, the H. pylori ABO blood group antigen binding adhesin. Science 305:519-525.

Avent ND (2001) Molecular biology of the Rh blood group system. J Pediatr Hematol Oncol 23(6):394-402.

Berger SA, Young NA, Edberg SC (1989) Relationship between infectious diseases and human blood type. Eur J Clin Micro Infect Dis 8(8):681-689.

Blajchman (2005) Landmark studies that have changed the practice of transfusion medicine. Transfusion 45(9):1523-30

Cartron JP, Colin Y (2001) Structural and functional diversity of blood group antigens. Transfus Clin Biol 8(3):163-99.

Greenwalt TJ (2005) Antibodies, antigens, and anticoagulants: a historical review of a lifetime in transfusion medicine-the Landsteiner Lecture 2004. Transfusion 45(9):1531-9.

Hadley TJ, Peiper SC (1997) From malaria to chemokine receptor: the emerging physiologic role of the Duffy blood group antigen. Blood 89(9):3077-91

Marshall BJ, Warren JR (1984) Unidentified curved bacilli in the stomach of patients with gastritis and peptic ulceration. Lancet 1(8390):1311-5.

Moore P (2003) Blood and justice: the seventeenth-century Parisian doctor who made blood transfusion history. John Wiley Chichester.

Rios M, Bianco C (2000) The role of blood group antigens in infectious diseases. Semin Hematol 37(2):177-85.

Scott ML (2001) Monoclonal anti-D for immunoprophylaxis. Vox Sang 81(4):213-8.

Speiser P (1975). Landsteiner K. English translation by Richard Rickett. Varlag Brüder Hollinek, Wien.

Chapter 6: Inhospitable Red Cells

Boyd JH, Watkins AR, Price CL, Fleming F, DeBaun MR (2005) Inadequate community knowledge about sickle cell disease among African-American women. J Natl Med Assoc. 97(1):62-7.

Dean M, Carrington M, O'Brien SJ (2002) Balanced polymorphism selected by genetic versus infectious human disease. Ann Rev Genomics Human Genetics 3:263-292.

de Montalembert M, Guilloud-Bataille M, Ducros A, Galacteros F, Girot R, Herve C, Maier-Redelsperger M, Feingold J (1996) Implications of prenatal diagnosis of sickle cell disease. Genet Couns7(1):9-15.

Durosinmi MA, Odebiyi AI, Akinola NO, Adediran LA, Aken'Ova Y, Okunade MA, Halim NK, Onwukeme KE, Olatunji PO, Adegoroye DE (1997) Acceptability of prenatal diagnosis of sickle cell anaemia by a sample of the Nigerian population. Afr J Med Sci 26(1-2):55-8.

Fixler J, Styles L (2002) Sickle cell disease Pediatr Clin North Am.49(6):1193-210.

Freimuth VS, Quinn SC, Thomas SB, Cole G, Zook E, Duncan T (2001) African Americans' views on research and the Tuskegee Syphilis Study. Soc Sci Med 52(5):797-808.

Graham R Sergeant and Beryl E Sergeant (2001) Sickle Cell Disease. Third edition. Oxford University Press, New York, NY.

Mason F "The Science and Humanism of Linus Pauling" (1901-1994)," February 1997, on the Web at http://osulibrary.orst.edu/specialcollections/coll/pauling/document-mason5.html

Modell B, Petrou M, Layton M, Varnavides L, Slater C, Ward RH, Rodeck C, Nicolaides K, Gibbons S, Fitches A, Old J (1997) Audit of prenatal diagnosis for haemoglobin disorders in the United Kingdom: the first 20 years. British Med J 315(7111):779-84.

Old JM (2003) Screening and genetic diagnosis of haemoglobin disorders. Blood Rev 17(1):43-53.

Richer J, Chudley AE (2005) The hemoglobinopathies and malaria. Clin Genet 68(4):332-6.

Rowley PT, Loader S, Sutera CJ, Walden M, Kozyra A (1991) Prenatal screening for hemoglobinopathies. III. Applicability of the health belief model. Am J Hum Genet 48(3):452-9.

Stuart MJ, Nagel RL (2004) Sickle-cell disease. Lancet364(9442):1343-60.

Vrettou C, Traeger-Synodinos J, Tzetis M, Palmer G, Sofocleous C, Kanavakis E (2004) Real-time PCR for single-cell genotyping in sickle cell and thalassemia syndromes as a rapid, accurate, reliable, and widely applicable protocol for preimplantation genetic diagnosis. Hum Mutat 23(5):513-21.

Weatherall DJ, Clegg JB (1981). The Thalassemia Syndromes. Third edition. Blackwell Scientific Publication. London, UK.

Chapter 7: Tainted Water

Bennett J, Ma C, Traverso H, Agha SB, Boring J (1999) Neonatal tetanus associated with topical umbilical ghee: covert role of cow dung. International Journal of Epidemiology 25:879-884.

Hirschhorn N (1982) Oral rehydration therapy for diarrhea in children-a basic primer. Nutr Rev 40(4):97-104.

Katz SL (2004) From culture to vaccine–Salk and Sabin. N Engl J Med 351(15):1485-7.

Whitfield J (2003) Gut reaction-News Feature Nature 423:583-584

Chapter 8: The Cholera Morbus

Adams GW (1961). Doctors in Blue: The Medical History of the Union Army in the Civil War, Collier Books New York.

Felsenfeld O (1967). The Cholera Problem. Warren H. Green, Inc. St Louis, Missouri

Gabriel SE, Brigman KN, Koller BH, Boucher RC, Stutts MJ (1994) Cystic fibrosis heterozygote resistance to cholera toxin in the cystic fibrosis mouse model. Science 266:107-109.

Rosenberg C. E. (Edi) (1972) The Cholera Bulletin–1832. Medicine and Society in America, Arno press and the New York Times, New York.

Sack DA, Sack RB, Nair GB, Siddique AK (2004) Cholera. Lancet 17;363(9404):223-33.

Thompson FL, Iida T, Swings J (2004) Biodiversity of vibrios. Microbiol Mol Biol Rev 68(3):403-31.

Chapter 9: The Burning Fever

Dubos RJ (1950) Louis Pasteur: Free lance of science by Little, Brown and Company. Boston.

Lax E. (2004) The mold in Dr. Florey's coat. Henry Holt and Co. NY, NY

Leavitt JW (1996) Typhoid Mary Beacon Press, Boston MA

Leinders-Zufall et al (2004) MHC class I peptides as chemosensory signals in the vomeronasal organ. Science 306:1033-1037.

Nicolle J (1961). Louis Pasteur: A master of scientific inquiry. Hutchinson and Co. London

Parry CM, Hien TT, Dougan G, White NJ, Farrar J (2002) Typhoid fever. The New England Journal of Medicine 347:1770-1781.

Ziporyn T (1988) Disease in the popular American press. The case of diphtheria, typhoid fever, and syphilis 1870-1920. Greenwood press New York, Westport CT, London. 1988.

Chapter 10: The Salty Sweat Disease

Cuthbert AW, Halstead J, Ratcliff R, Colledge WH, Evans MJ (1995) The genetic advantage hypothesis in cystic fibrosis heterozygotes: a murine study. J Physiology 482.2:449-454.

Egan et al (2004) Curcumin, a major constituent of turmeric, corrects cystic fibrosis defects. Science 304:600-602

Kerem B, et al (1989) Identification of the cystic fibrosis gene: genetic analysis. Science 245:1073-1080.

Marks G (1989) The cystic fibrosis gene is found. Science 245:923-925.

Pier GB (2000) Role of cystic fibrosis transmembrane conductance regulator in innate immunity to Pseudomonas aeruginosa infections. PNAS 97(16) 8822-8828.

Ratjen F, Döreg G (2003) Cystic Fibrosis. Lancet 361:681-689.

Riordan JR, et al (1989) Identification of the cystic fibrosis gene: cloning and characterization of complementary DNA. Science 245:1066-1073.

Roberts L (1988) The race for the cystic fibrosis gene. Science 240: 141-144.

Roberts L (1988) Race for cystic fibrosis gene nears end. Science 240:282-285.

Rommens JM, et al (1989) Identification of the cystic fibrosis gene: chromosome walking and jumping. Science 245:1059-1065.

Wilschanski M, et al (2003) Gentamycin-induced correction of CFTR function in patients with cystic fibrosis and *CFTR* stop mutations. New England Journal of Medicine 349:1433-1441.

Chapter 11: The Super Mutation

Becker Y (2005) The Molecular Mechanism of Human Resistance to HIV-1 Infection in Persistently Infected Individuals-A Review, Hypothesis and Implications. Virus Genes 31(1):113-9.

Dragic T, Litwin V, Allaway GP, Martin SR, Huang Y, Nagashima KA, Cayanan C, Maddon PJ, Koup RA, Moore JP, Paxton WA (1996) HIV-1 entry into CD4+ cells is mediated by the chemokine receptor CC-CKR-5. Nature 381(6584):667-73.

Liu R, Paxton WA, Choe S, Ceradini D, Martin SR, Horuk R, MacDonald ME, Stuhlmann H, Koup RA, Landau NR (1996) Homozygous defect in HIV-1 coreceptor accounts for resistance of some multiply-exposed individuals to HIV-1 infection. Cell 86(3):367-77.

Mills SG, DeMartino JA (2004) Chemokine receptor-directed agents as novel anti-HIV-1 therapies. Curr Top Med Chem 4(10):1017-33.

Paxton WA, Martin SR, Tse D, O'Brien TR, Skurnick J, VanDevanter NL, Padian N, Braun JF, Kotler DP, Wolinsky SM, Koup RA (1996) Relative resistance to HIV-1 infection of CD4 lymphocytes from persons who remain uninfected despite multiple high-risk sexual exposure. Nature Med 2(4):412-7.

Chapter 12: Natural-Born Killer

Bibel DJ, Chen TH (1976) Diagnosis of Plague: an analysis of the Yersin-Kitasato controversy. Bacteriological Reviews. 40(3):633-651.

Timothy L Bratton. The identity of the Plague of Justinian Trans Stud Coll Physicians Phila 1981 3(2):113-24.

Cantor F (1984) In the wake of the plague: the blackdeath and the world it made. 1984. The Free Press. New York, London, Toronto, Sydney, Singapore.

Carmichael AG (1986) Plague and the poor in Renaissance Florence Cambridge University Press. Cambridge England1986

Cornelis GR (2000) Molecular and cell biology aspects of plague. Proc Natl Acad Sci 97(16):8778-8783.

Galvani AP, Slatkin M (2003) Evaluating plague and smallpox as historic selective pressures for the CCR5-Δ32 HIV-resistance allele. Proc Natl Acad Sci 100(25):15276-15279.

Gross L (1995) How the plague bacillus and its transmission through fleas were discovered: reminiscences from my years at the Pasteur Institute in Paris. Proc Natl Acad Sci 92:7609-7611.

Kitasato S (1894) The bacillus of bubonic plague. Lancet Aug 25 pages 428-430

Loeb M (2004) A shot in the arm. Nature 431;892-893.

Edward Marriott (2003) Plague: A story of science, rivalry and the scourge that won't go away. Henry Holt and Co.

Mecsas J et al (2004). CCR5 mutation and plague protection. Nature 427:606

Parkhill J, et al (2003) Genome sequence of *Yersinia pestis*, the causative agent of plague (2003) Nature 413:523-527.

Perry RD, Fetherston JD (1997) Yersinia pestis-etiologic agent of plague. Clinical Microbiology Reviews 10(1):35-66.

Rollins SE, Rollins SM, Ryan ET (2003) *Yersinia pestis* and the plague. Am J Clin Pathol 119:S78-S85.

Stephens JC et al (1998) Dating the origin of the CCR5-Δ32 AIDS-resistance allele by coalescence of haplotypes. American Journal Human Genetics 62:1507-1515.

Solomon T (1995) Alexandre Yersin and the plague bacillus. Journal of Tropical Medicine and Hygiene 98:209-212.

Wren BW (2003) The yersiniae-A model genus to study the rapid evolution of bacterial pathogens. Nature Reviews Microbiology 1:55-64

Chapter 13: The King of Terrors

Aldrete JA (2004) Smallpox vaccination in the early 19th century using live virus carriers: the travels of Francisco Xavier de Balmis. Southern Medical Journal 97(6):375-378.

Bourzac K (2002) Smallpox: Historical review of a potential bioweapon. Journal of Young Investigators, Issue 3, volume 6.

Barquet N, Domingo P (1997) Smallpox: the triumph over the most terrible of the ministers of death. Ann Intern Med. 1997 Oct 15;127(8 Pt 1):635-42.

Cantor E (2001) In the wake of the plague. The Free Press. NY, NY.

Cunha BA (2004) Smallpox and measles: historical aspects and clinical differentiation. Infectious Disease Clinics of North America 18:79-100.

Eyler JM (2003) Smallpox in history: The birth, death, and impact of a dread disease. J Lab Clin Med. 142:216-220.

Lalani AS, et al (1999) Use of chemokine receptors by poxviruses. Science 286:1968-1971.

McCullough D (2002) John Adams Touchstone NY, NY

Patterson KB, Runge T (2002) Smallpox and the Native American. Am J Med Sci 323(4):216-222.

Spier RE (2002) Perception of risk of vaccine adverse events: a historical perspective. Vaccine 20:S78-S84.

Tramont EC (2004) The impact of syphilis on humankind. Infectious Disease Clinics of North America 18:101-110.

Chapter 14: Gene Splicers

Barber CG (2004) CCR5 antagonists for the treatment of HIV. Curr Opin Investig Drugs. 5(8):851-61.

Birge AE (1988) Bacterial and bacteriophage genetics. Springer-Veralg. New York.

de Silva E, Stumpf MP (2004) HIV and the CCR5-Delta32 resistance allele. FEMS Microbiol Lett 241(1):1-12.

Lederberg J (1986) Forty years of genetic recombination in bacteria. A fortieth anniversary reminiscence. Nature 324(6098):627-8

Lin ECC, Goldstein R, Syvanen M (1984) Bacteria, Plasmids and Phages. Harvard University Press. Cambridge, Massachusetts and London, England.

Merril CR, Scholl D, Adhya SL (2003) The prospect for bacteriophage therapy in Western medicine. Nature Reviews (drug discovery) 2:489-497.

Skuse DH et al (1997) Evidence from Turner's syndrome of an imprinted X-linked locus affecting cognitive function. Nature 387:705-708.

Nathans D, Smith HO (1975) Restriction endonucleases in the analysis and restructuring of DNA molecules. Annu Rev Biochem. 1975;44:273-93.

Thacker PD (2003) Set a microbe to kill a microbe. JAMA 290:3183-3185.

Chapter 15: Reversing the Code

Baltimore D (1970) RNA-dependant DNA polymerase in virions of RNA tumour viruses. Nature. 226(252):1209-11.

Bishop JM (1983) Oncogenes. Scientific American 246: 80-92.

Bernstein J (1998) Science, Fraud & the Baltimore Case. Commentary Magazine, Dec, 1998,

Der CJ, Krontiris TG, Cooper GM (1982) Transforming genes of human bladder and lung carcinoma cell lines are homologous to the rat genes of Harvey and Kirsten Sarcoma virus. Proceedings of the National Academy of Sciences 79(11):3637-3640.

Kevles DJ (1988) The Baltimore Case: A Trial of Politics, Science, and Character First edition. W.W. Norton & Company;

Land H, Parada LF, Weinberg RA (1983) Tumorigenic conversion of primary embryo fibroblasts requires at least two cooperating oncogenes. Nature 304:596-602.

Spalding BJ (1999) It's a wonderful life: The vindication of Nobel Laureate David Baltimore. 12/22/99. http://www.biospace.com/articles.

Temin HM, Mizutani S (1970) RNA-dependant DNA polymerase in virions of Rous sarcoma virus. Nature 226(252):1211-3.

Varmus H (1988) Retroviruses. Science 240:1427-1435.

Varmus HE (1990) Nobel lecture. Retroviruses and oncogenes. I. Biosci Rep. 10(5):413-30.

Weinberg RA (1983) A Molecular Basis of Cancer. Scientific American 249:126-142.

Chapter 16: Battle of the Genomes

Collins F, Morgan M, Patrinos A (2003) The human genome project: lessons from large-scale biology. Science 300:286-290.

Gardner MJ, et al (2002) Genome sequence of the human malaria parasite *Plasmodium falciparum*. Nature 419:498-511.

Holt RA et al (2002) The genome sequence of the malaria mosquito *Anopheles gambiae*. Science 298:129-149.

Lander ES et al (2001) Initial sequencing and analysis of the human genome. Nature 409:860-921.

New York Times Magazine, (November 28, 1999). Sheryl Gay Stolberg. The Biotech Death of Jesse Gelsinger

Roberts L (2001) Controversy from the start. Science 291:1182-1185.

Venter JC et al (2001). The sequence of the human genome. Science 291:1304-1351.

Index